For my beloved D . . .

. . . and our beloved Oli

Edward Davey is a Project Director at the World Resources Institute and has worked for The Prince of Wales's International Sustainability Unit on global environmental issues, leading the organisation's work on forests and climate change, as well as for the Colombian presidency on Colombia's natural environment and the conservation of the Amazon. He has spent time in Iraq and the Middle East for a humanitarian charity. *Given Half a Chance* is his first book.

With special thanks to:

Richard Davey
Ben Goldsmith
Hylton Murray-Philipson
Jessica & Adam Sweidan
Steve Wake

Given Half a Chance

TEN WAYS TO SAVE THE WORLD

Edward Davey

unbound

This edition first published in 2019

Unbound
6th Floor Mutual House, 70 Conduit Street, London W1S 2GF

www.unbound.com

Text design by Ellipsis, Glasgow

A CIP record for this book is available from the British Library

ISBN 978-1-78352-659-8 (paperback)
ISBN 978-1-78352-661-1 (ebook)

Printed and bound in Great Britain by Clays Ltd, Elcograf S.p.A.

1 3 5 7 9 8 6 4 2

Contents

The difference between what we do and what we are capable of doing would suffice to solve most of the world's problems

– Mahatma Gandhi

Oli

This is a book about the world in which my son Oli will grow up. I wrote the bulk of it in the months leading up to his birth at the Royal Free Hospital in London on Friday, 28 April 2017. I then returned to it during the year which followed, alongside all the joys and rigours of his first year of life.

My purpose is a simple one. I want to describe ten paths by which we can safeguard and protect this precious planet into which Oli has been born. My dream is of a world in which we have found a way to live in harmony with the natural environment on which we depend, and in so doing to live also more fully in peace with one another. In the book, I show how positive action in ten key areas could bring about a restored earth and a more hopeful future. These paths include transforming our cities, energy and waste systems; protecting our vital forests, soils, fresh water, ocean and biodiversity; addressing population growth, unhealthy diets and over-consumption; and pursuing global action to address these issues at a scale we have never seen before.

Preface

Each of the paths I describe is aimed first and foremost at addressing climate change and ecological decline. I do not set out how to deliver peace in the Middle East, end the terrible war in Syria, eliminate the risk of nuclear warfare, clean up nuclear waste, end poverty or address antimicrobial resistance. There are other existential challenges of our time which are beyond the scope of this book.

Nevertheless, my ten paths are intended to deliver a better world in ways which go far beyond the environment. Tackling climate change head on – by massively reducing our greenhouse gas emissions, as well as adapting well to such changes as are already inevitable – will significantly reduce the likelihood of today's levels of conflict and migration being further exacerbated by the increasing incidence of droughts, floods and natural resource scarcity. Protecting biodiversity could prove crucial to scientific and medical research, with huge benefits to global public health. Managing the world's rivers more effectively and equitably would reduce the prospect of conflict between states. Ensuring that the cities of the twenty-first century are green, well-planned and 'low carbon' would be beneficial for human lives in many ways – not least in terms of people's physical and mental health – as well as being vital for the climate and the environment. The circular economy – an economy based on innovation, renewables, recycling and zero waste – could generate many new and better jobs for people around the world, thereby playing an important part in ensuring a 'just transition' for workers who stand to lose out from the phasing out of today's energy systems. The set of actions and reforms described in my tenth path – in economics, politics, legal systems, the media, the faith community and elsewhere – could play a big role in healing the rifts, divisions and anger of our times.

Despite the scale of the environmental challenges before us, and the sheer speed with which we need to act to bring about the transformations described here, I am deeply hopeful that it is within humanity's power and intellect to bring into being the kind of world that I would wish to leave Oli. It is in our hands: we know with great clarity what needs to be done, and in many parts of the world, we are already doing it. But there is much more to be done: we are in a race against time, and so we will need to act with all the bravery, foresight, intelligence and commitment we can muster.

Fortunately, the overwhelming majority of the world's people are on our side: there is more that unites us than divides us, as the late British parliamentarian Jo Cox would say, as true for our globally shared environmental concern as anything else. When I met with the distinguished British doctor and UN leader, David Nabarro – one of the fifty or so interviewees whose thoughts inform this book – he referred to his 'absolute belief in the power of the human', and 'that out of 1,000 humans, 999 of those are working for the good of everybody and for the renewal of our planet'. It is easy to forget how much progress is being made to bring about a better world every day: one of my aims here is to draw attention to some inspiring examples of where we are getting it right, and to reflect on how we might build on these approaches and take them to scale.

This is not a false optimism: I am under no illusions that addressing the challenges I describe here will be easy. We have, without question, a mountain to climb, whether to keep climate change within safe bounds, to stop tropical forests being lost or to clean up the plastic in the ocean. One can but mourn, daily, the environmental impact we have already had, whether on the coral reefs, biodiversity, freshwater supplies or the world's soils. But I am also deeply persuaded by Mahatma Gandhi, when he said: 'The difference between what we do and what we are capable of doing would suffice to solve most of the world's problems.' It is within our grasp, now, to say that 'enough is enough' and to establish a new direction of travel for the world: a world in which we establish economies and societies in which we all live well, and comfortably, while at the same time profoundly respecting the earth and the natural environment. The time for hand-wringing is over; the time for purposeful action and solutions is upon us.

As Erik Solheim, the Executive Director of the United Nations Environment Programme, reminded me, humanity can change its path fast about a given wrong: the abolition of the slave trade, the Montreal Protocol (a global agreement to protect the stratospheric ozone layer) and the banning of smoking in restaurants are just a few examples of where the world has brought about a radical change to a situation within a very short space of time. Throughout the period I spent writing this book, barely a day went by without a city announcing a ban on plastic bags, a company committing to reducing its environmental impact or a community finding new ways to reduce its waste. And so, there is great hope for similarly fast global environmental action to occur to address the challenges described in this book. Indeed, there is a lot of evidence to suggest that we are on the cusp of a monumental transition, with more people concerned about the environment than ever before. The challenge is to capitalise on these high levels of concern and engagement to bring about real and lasting changes at scale.

Given Half a Chance is not intended to be a textbook, nor a comprehensive account of all the issues. Instead, I hope to provide the general reader with a snapshot of the state of the world, its principal environmental challenges and some of the most promising solutions, and to play a small role in galvanising greater collective action. Along with recounting some of my personal experiences working on the environment, my primary research method has been to seek out a number of the world's leading experts, advocates, campaigners, business people and moral leaders with a commitment to the environment, to ask for their guidance on each of the paths and to talk through examples of the kinds of changes under way which most inspire them.

Above all, I hope the book will inspire its readers to greater action, both in terms of the decisions they take in their own lives, and, above all, in clamouring for their leaders to undertake a more determined and far-sighted political response to bringing about these ten paths.

Preface

Ultimately, it is my view that we will need sustained political will at the highest level to turn the situation around; and this will only come about if *people* push for this to happen. Here, too, I am deeply hopeful.

INTRODUCTION

2015

A few months before Oli's birth, on a cold, wintry morning in January 2017, I went to speak to one of my great heroes, a man who has seen more of the natural world than perhaps anyone else, and someone with a very clear set of ideas about what we most need to do to put things right: Sir David Attenborough.

The sky that morning in south-west London was a bright blue, full of planes flying low and in quick succession into Heathrow. When I arrived at the train station, I asked the man at the gate to give me directions; he looked me up and down and asked if I was off to see David Attenborough. Somewhat bashfully, I said yes; he smiled and wished me well. Sir David is clearly as well-known locally as he is celebrated internationally. As I approached the house, I noticed a few parakeets and pigeons flying overhead in tandem: I wondered whether this was symbolic of a meeting of London's and the world's biodiversity, before putting such thoughts to one side, knocking on the door and being welcomed in by Sir David and his daughter Susan.

Introduction

To the backdrop of a roaring gas fire and a wonderful collection of books, music and artefacts, and over a cup of tea, I began by asking Sir David what two or three things he would focus on in an attempt to bring about a more sustainable world. His first overwhelming priority, he replied, is to address global warming: 'a global phenomenon which will cause damage to the entire globe, through all kinds of pervasive effects: addressing the earth's rising temperature and its climatic impact on the land and the ocean is our main project'. His solution was straightforward: a 'concentration on moving to renewables from our carbon-based economy. The sad and frustrating thing is of course that we know how to do it: we have all the technology, although it needs a certain amount of refinement, and once we know how to store and relay renewable energy, and renewables reach a price which undercuts the current price of oil and coal, the problem is answered: society will answer it itself.'

His second prescription was to step up the global effort, through education and access to voluntary reproductive health care, to lower and stabilise rates of population growth, although he was not sanguine about the prospects of success. 'Rising population is a huge problem,' he said; 'it will certainly go on rising whatever we do. It is every human being's fundamental human right you are talking about in this instance: whether we can influence that or not, I don't know, but the consequences of not doing so seem clear to me.'

His third priority is to bring about dramatically more sustainable cities, which he sees as a 'way of making life for some people in the city more attractive, more variable and more in tune with the natural world, as well as providing particular species with a home they might not have had otherwise'.

I explained that I aimed to put forward a hopeful case in this book as to how we can now safeguard the planet and put it on a better track, including through concerted action in these three areas, and asked him whether he was optimistic or pessimistic in light of all that he had seen in his lifetime. He responded: 'This is not a black and white question and it doesn't have a black and white answer. No doubt from some points of view the world is going to get worse. Humanity can survive, but it will not have as much fun as I have had and see the things I have seen. Nor will we be able to travel in the way that I used to travel, to go to some parts of the world with no way of communicating.' But he was clear that there is still so much to do and so much worth fighting for. He was particularly enthused by the unprecedentedly high numbers of people around the world who are concerned about the state of the planet and keen to make amends; this gives him great hope. He was also proud of the fact that many hundreds of young people wrote to him to declare their intentions to work in conservation, their careers having changed as a result of seeing his programmes.

Before I left, he signed a copy of his book *The Living Planet* to Oli, and Susan took a photograph of our encounter: a photo I will treasure, with a man of boundless intellect and enthusiasm, who has done more than perhaps anyone else to encourage humanity to love and to respect the natural world.

Before we set out what needs to be done to safeguard our planet, let us first take stock of where we are, and why I believe there are such grounds for hope. Some of my optimism heralds from the fact that in the year 2015 (not so very long ago, although it sometimes feels a world away), all 193 of the UN's member states signed up to two inspiring and mutually reinforcing new global agreements, the UN Sustainable Development Goals (or 'Global Goals', as I shall refer to them) and the Paris Agreement on climate change. Both of these agreements have also been signed up to by thousands of companies, regions, cities, faith groups, trade unions and civil society organisations. In early 2015, in the 800th anniversary year of the Magna Carta, the Prince of Wales referred to the two agreements as a 'Magna Carta for the earth, and humanity's relationship with it'. Together, these two sets of goals – if implemented in full, and with commitment and sincerity by all who have signed up to them – could put the world on a radically different and better course.

The 'Global Goals' were the first of the two milestones agreed at a meeting at the UN Headquarters in New York in September 2015. There are 17 goals, reproduced below, for the world to achieve by 2030, backed up by 169 more detailed targets:

1. **No Poverty** – End poverty in all its forms everywhere
2. **Zero Hunger** – End hunger, achieve food security and improved nutrition, and promote sustainable agriculture
3. **Good Health and Well-being** – Ensure healthy lives and promote well-being for all at all ages
4. **Quality Education** – Ensure inclusive and equitable quality education and promote lifelong learning opportunities for all
5. **Gender Equality** – Achieve gender equality and empower all women and girls
6. **Clean Water and Sanitation** – Ensure availability and sustainable management of water and sanitation for all
7. **Affordable and Clean Energy** – Ensure access to affordable, reliable, sustainable and clean energy for all
8. **Decent Work and Economic Growth** – Promote sustained, inclusive and sustainable economic growth, full and productive employment and decent work for all
9. **Industry, Innovation and Infrastructure** – Build resilient infrastructure, promote inclusive and sustainable industrialisation and foster innovation
10. **Reduced Inequalities** – Reduce inequality within and among countries
11. **Sustainable Cities and Communities** – Make cities and human settlements inclusive, safe, resilient and sustainable
12. **Responsible Consumption and Production** – Ensure sustainable consumption and production patterns

13. **Climate Action** – Take urgent action to combat climate change and its impacts
14. **Life Below Water** – Conserve and sustainably use the oceans, seas and marine resources for sustainable development
15. **Life on Land** – Protect, restore and promote sustainable use of terrestrial ecosystems, sustainably manage forests, combat desertification, and halt and reverse land degradation, and halt biodiversity loss
16. **Peace, Justice and Strong Institutions** – Promote peaceful and inclusive societies for sustainable development, provide access to justice for all and build effective, accountable and inclusive institutions at all levels
17. **Partnerships for the Goals** – Strengthen the means of implementation and revitalise the global partnership for sustainable development

I find the Global Goals inspiring for three reasons. First, I believe they can be used to hold governments to account. In the year 2000, at university, I remember celebrating the creation of the original UN Millennium Development Goals (the 'MDGs', a precursor to the Global Goals) and participating in the Jubilee 'Drop the Debt' movement which captured the public imagination at the time. On the basis of the MDGs, we lobbied Parliament for Britain to be a development leader and to meet the UN goal of giving 0.7 per cent of its GDP (gross domestic product) to overseas development. On leaving university, fired up with idealism about ending poverty and addressing climate change, I volunteered for the 2005 'Make Poverty History' campaign, intended to catalyse action and finance towards the fulfilment of the MDGs. We marched to Parliament, again, and heard Nelson Mandela speak in Trafalgar Square. The MDGs provided a rallying cry for a whole civic movement in the UK – including churches, trade unions and student groups – committed to poverty alleviation in the world. They also inspired governments around the world to act to address malaria, reduce poverty and tackle HIV/AIDS.

My second reason is that I find the Global Goals much more visionary on the importance of environmental sustainability than the original MDGs were (only one of their original eight goals was environmental, and it appeared to be something of an afterthought). At least half of the Global Goals, by contrast, are fully focused on the environmental challenges described in this book. And so while some, including my cousin Oliver Phillips (Professor of Tropical Ecology at the University of Leeds), would say that these goals are still not environmentally ambitious enough, but instead primarily 'anthropocentric' and more focused on the rights of humans than of the earth itself, I nevertheless believe they find a much better balance than before. (The Colombian Government, for which I used to work, played a significant role in achieving this balance.) The Global Goals are also avowedly universal in scope, as relevant for rich countries as they are for poor ones, with strong messages on inequality, over-consumption and climate action.

Thirdly, I have seen how the Global Goals have already given rise to much positive effort, funding and action across the world: a number of countries have presidential commissions to implement the goals; some businesses, such as Unilever, have reoriented their business plans to fulfil them; and scientists and academia have used them as a frame for much helpful analysis, research and study on how sustainable development might be achieved. And so, while the Global Goals have not yet acquired the global relevance and political attention they deserve, it is certainly a framework to uphold and to implement; a means of holding a mirror up to the state of humanity and the planet and to see whether we are on track to creating 'the future we want', or way off course.

A few months after the Global Goals were agreed, 151 world leaders congregated in Paris on 30 November 2015 for the opening day of the 21st UN Climate Change Conference: the greatest ever assembly of presidents and prime ministers in one room on the same day, on any subject. I happened to be a privileged bystander in the room that day, as the Prince of Wales had been invited to give the opening address to the talks, and enjoyed seeing Obama, Modi, Xi Jinping, Merkel, Hollande, Cameron and others at close quarters.

Two weeks of arduous negotiations followed the leaders' opening speeches, which set the bar high, and to much fanfare and celebration the Paris Agreement was adopted on 12 December 2015. The Agreement commits the international community to a set of policies and actions which would keep global temperature rise to under 2 °C (3.6 °F), and ideally 1.5 °C above pre-industrial levels. This latter number is particularly iconic, as a change of over 1.5 °C puts many of the world's small island states, as well as coastal cities, coral reefs and other areas of great value under existential risk. To reach 1.5 °C, let alone 2 °C, will require a radical trans-formation of the global economy, fast. It is scientifically possible, but hugely ambitious. The sum total of coun-tries' current national contributions to meeting the Paris Agreement, assuming they are fully implemented, would lead to at least a 3.6 °C rise, which would be calamitous for countries and communities around the world. According to the UK Met Office, a 4 °C rise in global temperatures would have catastrophic effects, including the potential 'dieback' of the Amazon rainforest, a reduction in global agricultural yields, reduced fresh-water supply for at least 1 billion people and twice as frequent drought events in the Mediterranean Basin, South-east Asia and Southern Africa.

The Paris Agreement nevertheless sends a strong and unequivocal signal to the financial community, investors and markets that the world has a long-term commitment to addressing climate change. It refers to the need to 'balance carbon sources and sinks', which significantly raises the profile of protecting forests and investing in more sustainable agricultural practices, given the role forests and soils play as vital storing houses for carbon. It calls for fossil fuel subsidy reform, thereby tackling the stark fact that the world's countries continue to subsidise fossil fuel extraction to the tune of US$5.3 trillion per year. It also commits countries to undertake 'long-term decarbonisation pathways': clear, scientifically rigorous and cross-governmental long-term plans to cut their greenhouse gas emissions. It makes a solemn vow to provide proper funding for climate action, above all in the form of wealthier countries supporting poorer and more vulnerable countries to pay for the 'loss and damage' which accrues to them as a result of climate change. Finally, the Agreement also dictates that the world should not emit more greenhouse gas emissions than the planet can absorb, with the aim of achieving 'zero net' carbon growth, or 'climate neutrality', by the second half of this century. In so doing, the Agreement simultaneously recognises that the world's current efforts are insufficient, while creating an unequivocal long-term goal to which we can and must aspire. At a time of significant failures of the multilateral system, for example over Syria, the Paris Agreement represented an important example of multilateral diplomatic success. There was much to celebrate at the end of 2015.

In addition to the intelligent diplomacy and careful preparation of the French hosts, the success of the Paris talks was also attributable to Christiana Figueres, the energetic and ebullient climate diplomat who had been Executive Secretary of the UN Framework Convention on Climate Change since 2010, after the failure of the infamous UN Copenhagen Climate Summit in late 2009, where talks at the eleventh hour between the US and

China had broken down. Christiana and I met in London for a cup of tea overlooking the River Thames, and I began by asking her for her take on how success in Paris had come about.

She said that 'a number of conditions had changed since the trauma of Copenhagen, when no one, myself included, had ever imagined that the world would one day pick itself up and establish a new agreement'. The first was the economics of renewable energy; a technical solution to climate change appeared in 2015 so much more possible than it had been in 2009. By 2015, 'the cost of renewables had fallen as much as 80 per cent, and there was extraordinary evidence from over 40 jurisdictions that renewables were at least as cost effective, if not cheaper than, fossil fuels'. Also in the six-year gap, many more countries had been negatively impacted by climate change. The injustice was undeniable: the people and countries most vulnerable to climate change, such as Haiti, bore 'absolutely no responsibility for it'. Such countries 'would never be able to pursue the Global Goals because they would have to rebuild constantly, the social injustice of which was absolutely unbearable', Christiana said.

Our conversation continued. Where Copenhagen had sought to establish a 'top down' climate agreement, 'baking a cake, cutting it into slices, and then parsing those slices out' to each country on the basis of their respective responsibilities for the problem, Paris had represented a 'paradigm shift to a model of future responsibility on the shoulders of all', which was built on the basis of what each country was willing to bring to the table. While historical responsibilities for the bulk of emissions remained important, so too was an understanding that future emissions were likely to be greatest in the 'global South': the situation therefore called for a future in which every country would play its fair part in addressing the problem. Also, instead of pitting global needs against the national interest, in which the national invariably wins, it was now widely understood that addressing climate change was in a nation's best interest, and good for a country's economic growth; climate action therefore made perfect sense. And while Christiana recognised that it was clear that the sum total of the commitments made to date would not achieve the 2 °C limit, she was pleased that countries had committed to a long-term goal: the journey had begun.

The year 2015 was, as we have seen, a high point for global environmental action. Despite a fractured world, the international community had somehow managed to put its differences aside and broker these two ambitious agreements. It was an exciting time for someone with an interest in sustainable development and climate change: at times, indeed, ecstatic. President Barack Obama was in power in the US and was doing everything that he could to act on climate change, marine conservation and much else besides. He and President Xi Jinping had brokered a strong US–China deal on climate which was another key linchpin of the Paris Agreement and near guarantor of a positive outcome. The European Union, once at the forefront of international climate action, was now if anything something of a laggard. India and other key emerging economies also appeared keen to push for success. Prime Minister Justin Trudeau, new to office in Canada, made bold statements about climate change and the environment (although his subsequent record was much more mixed). Chancellor Angela Merkel, a scientist with a track record of understanding climate change since the late 1980s, remained resolved to ensure strong German domestic and global policy leadership. Pope Francis set out to galvanise climate action across the world's more than one billion Catholics, with a radical Papal Encyclical *On Care for our Common Home*. In Colombia, President Juan Manuel Santos had become an articulate advocate of the Global Goals and climate action, linking both to the peace process under way in his country. And in the UK, despite broader social and political change, there appeared still to be fairly strong consensus on the need for climate action and support for international development.

And then 2016 and 2017 happened, with a series of shifts in global politics, culminating in Brexit and the election of President Donald Trump. The liberal world order appeared to have been turned on its head, replaced by populist uprisings and real, profound disquiet across societies and regions of the world. The new mood was more autocratic, nationalist, authoritarian and aggressive, across Europe, the US, Russia, Brazil, Turkey and elsewhere. In Syria, the tragic conflict continued to deteriorate, as did the prospects of peace in the Middle East, while the prospect of famine stalked South

Sudan, Nigeria, Yemen and Somalia. North Korea threatened nuclear war. France toyed with electing Marine Le Pen. Thousands of refugees continued to cross the Mediterranean in precarious boats. A series of terrorist attacks took place in Germany, Sweden, France and London. And then, sure enough, President Trump announced that the US would withdraw from the Paris Agreement. It has been a deeply troubling time, and one in which the lofty agreements of 2015 feel both light-years away and distant to the wants and desires of many electorates.

Rather than write these agreements off, however, I would argue that they are more needed than ever before. Many of the most acute grievances people feel about the direction in which the world is heading would be powerfully assuaged by a wholehearted effort to meet the Global Goals, in particular their commitment to dignified work, education, reduced inequality and better infrastructure. To reach the Global Goals and the Paris Agreement will also require a much better political system – more authentic, more responsive to people's needs and more community-oriented and bottom-up – which would go some considerable way to re-involving people who feel left behind by politics and cynical about public policy. To meet the Paris Agreement would be to ensure whole countries remain viable and inhabitable, capable of withstanding a volatile climate and ensuring a dignified life for their citizens. The whole world order ultimately hinges on the Paris Agreement's goals being met: it is the single most important international security challenge of the century to come. And so all those who have invested so much in delivering these two agreements at the end of 2015 must not cease from explaining their importance to countries and communities today. The stakes are too high.

In these uncertain times, and whatever our disagreements, it is clear that the future of the world rests in our hands more than ever before. The decisions we take in the coming years will define whether we continue to live in a habitable world or not: whether we can live in peace with one another, as well as in broader harmony with the natural environment which surrounds us. We must tread carefully, and act with an overwhelming duty of care – to ourselves, to future generations and to the planet itself – to steer human behaviour and policies on a path which leads to peace, health and survival for us all. The agreements of 2015 pointed to what is possible: they serve as a guiding light. Our collective challenge is to bring them into being, fast, using every tool at our disposal to do so.

Renewables

The world is changing fast, with the cost of renewables plummeting and unprecedented levels of our energy needs now being met by wind and solar power. In the Nevada desert in the US, the 'Crescent Dunes Solar Energy Facility' opened in 2015. It occupies 1,670 acres of desert, with 10,347 billboard-sized mirrors, each 37 feet wide and 24 feet tall, and generates enough electricity for 75,000 homes. The thermal energy captured from the sun is stored in molten salt, which is pumped through a heat exchanger to turn water into steam that spins a turbine to generate electricity. Across the state of Nevada, around 360,000 homes are now powered by solar energy, 7.45 per cent of the state's electricity supply. Solar energy is also responsible for the existence of 16 manufacturers, 77 installers and 8,371 jobs in Nevada. A new project, currently under construction, will power another million homes. This is a fully integrated energy-storage technology, with distributed generation and storage, driven by the world's first hybrid geothermal solar photovoltaic thermal power plants.

Nevada is committed to 25 per cent of its retail electricity sales coming from renewables by 2025.

There are many more similarly exciting initiatives worldwide attempting to transform the world's energy system to a 'low carbon', renewable one. Across the US, and at variance with President Trump's rhetoric, some 3 million jobs are currently generated by clean energy, of which 260,000 are in the solar industry. In the US, it cost $96 per watt for a solar module in the mid-1970s; today, it's 68 cents. So strong is the renewable offer that Google, alongside other major multinational companies, now offsets 100 per cent of its energy usage with either wind or solar power. Chile has recently tendered six major new energy contracts; all the winning bids have been for renewables. South Africa, which relies on coal for 80 per cent of its electricity, has just approved $4.7 billion worth of investment in solar and wind projects. In Morocco, the percentage of the country's power generated from concentrated solar continues to increase, while Iceland is diversifying its energy supply to include low-heat geothermal.

Advances in battery storage mean that solar energy can be stored and fed into the grid when required, night or day. China is also making huge advances in domestic renewable energy. Where the central government previously had high targets for coal use, it is now closing coal power stations every week, becoming instead the world's foremost producer of wind, solar and battery storage (although it continues to fund coal power stations in other countries, such as Pakistan and Kenya). As a result, air quality has dramatically improved in a number of Chinese cities, as I saw for myself on a recent visit to Beijing where the city was blessed with bright blue skies. The cost of solar panels around the world has dropped significantly as a result of Chinese production, enabling countries to invest in massive infrastructure programmes based around energy efficiency, smart grids and proper interconnectors, and to think about energy in a smarter way.

Renewable energy solutions can be on a smaller, decentralised scale too, generating income as well as energy for local communities. In the UK, the inspiring Westmill Solar Co-operative in Oxfordshire consists of 30 acres of over 20,000 solar panels, and sells surplus energy back to the national grid. Westmill is the UK's first and the world's largest co-operatively run, community-owned solar farm. To walk through its installations, set in glorious countryside, is to witness a different possible future for Britain and the world.

I spoke with Sir David King, then the UK Foreign Secretary's Special Representative for Climate Change, to find out more about the role of innovation in delivering a renewable revolution. Sir David formerly served as the government's Chief Scientific Advisor and while at the Foreign Office had formed a plan with the London School of Economics Professor Richard Layard, the Global Apollo Project, to stimulate global investment in research and development for renewable energy. They argued that all it would take to transform the global energy system would be for renewables to be consistently cheaper than fossil fuels.

India's Prime Minister Narendra Modi signed his government up to play a central role in this project, provided it be renamed 'Mission Innovation'. David Attenborough became an enthusiastic advocate and when President Obama interviewed Attenborough for his ninetieth birthday – as part of Obama's efforts to convince the American people to accept climate action during the build-up to Paris, Attenborough's programmes being widely celebrated across the US – the pair discussed this project. President Obama was convinced of its value and the US joined Mission Innovation in time for the Paris climate talks, as did some 23 other governments. The pledge was simple: each signatory had to double its investments in renewable energy research and innovation by 2020. Bill Gates came on board investing $1 billion, and encouraged 23 other philanthropists to do the same. Both Mission Innovation and its sister initiative, Breakthrough Energy, set the tone at the outset of the Paris negotiations.

Mission Innovation invests in seven types of renewables research and development. These include smart grids, which are power grids run on renewable energy sources, able dynamically to adjust supply and demand in order to handle the intermittency of solar and wind power. The conversion of sunlight into storable solar fuels, such as hydrogen, is another priority, as are finding new affordable ways of heating and cooling buildings.

David King explained that one of Mission Innovation's most exciting ventures, Heindl Energie, came from a German mining engineer who had developed a simple, small-scale hydropower energy system with huge potential, 100 of which, placed at strategic locations, would meet all of the UK's energy needs. If every home were connected to a smart grid, it would mean appliances like a washing machine would only come on when the grid had enough renewable energy to power them. Elon Musk has also recently launched – as part of Tesla's takeover of the renewable firm SolarCity – a new design of camouflaged solar rooftop tiles, economically priced, which are gaining widespread take-up in the US.

The point of Mission Innovation is to let a thousand flowers bloom and to take risks. It is now scouring the world for viable investments, and continuing to build momentum: in March 2018, the UK and Saudi Arabia committed to work together to share technical knowledge and expertise on clean energy, and ministers from across participating countries met in May 2018 to call for an independent international fund to finance energy technology projects. Impressive investments are being made in battery storage technology, electrifying public transport and improving low carbon alternatives to heating, such as heat pumps. The race is on.

The pace of change is, however, by no means fast enough: levels of carbon dioxide in the atmosphere continue to rise month by month. May 2018 saw a record high of 411 parts per million (according to the Keeling Curve measurement series kept at the Mauna Loa observatory in Hawaii, recording the ratio of one gas to another), when the safe limit for the climate is widely held to be 350 parts per million. The world faces dangerous climate tipping points, such as the melting of the Arctic permafrost, which would lead to huge additional methane emissions, or the drying out of large areas of the Amazon rainforest, which would cause it to release more carbon than it absorbs.

The evidence of a changing climate is all around us: each year breaks new temperature records, while the world is beset with floods, droughts, melting ice caps and an ever more variable climate. The world is still a long way off a global emissions pathway that would stabilise climate change and meet the goals set in Paris. At the time of writing, current political commitments made by countries will reduce overall emissions by no more than a third of the levels required by 2030 to avert the risk of 'run-away' climate change.

Despite the rapid and widespread adoption of renewable technologies, and their plummeting costs, much more needs to be done. The current commitments of countries and companies to further coal, oil and gas exploration, if realised, will be sure to take us significantly beyond the temperature threshold agreed in Paris. The majority of the world's institutional investments and pension funds, as well as our government bonds, sovereign wealth funds and treasuries, are intimately tied up with the core investments and assumptions of a fossil fuel industry which falsely assumes it will exist in perpetuity. Louise Rouse, the articulate campaigner for 'Share Action' on fossil fuels, described to me the broader societal inertia which in part explains the difficulty in undertaking such a rapid transition. 'The fossil fuel industry has found a way of positioning itself at the heart of nations' cultural and academic life,' she explained: 'through its sponsorship of sports competitions, art exhibitions and university science departments, we find it hard to contemplate a world without oil and gas.'

But if we are to be serious about addressing climate change, the vast majority of the world's remaining fossil fuel reserves cannot in any reasonable scenario be exploited; they should instead be considered untouchable, or 'stranded assets', to borrow a phrase from the Oxford academic, Ben Caldecott. Globally, key players are becoming more aware of the extent of our over-reliance on fossil fuels; leaders in the financial community – including the Governor of the Bank of England, Mark Carney, and Managing Director of the International Monetary Fund, Christine Lagarde – have been increasingly eloquent and outspoken on this subject. Investors are now even more likely to question the oil majors' business models, ask tough questions, encourage proactive investment in renewable energy and divest from companies not committed to a growth model consistent with the Paris Agreement. But there is still a long way to go.

To understand the scale of the energy transformation required, I went to interview my friend John Sauven,

Executive Director of Greenpeace UK, a man of moral integrity and an inveterate campaigner on climate change, tropical forests and the marine environment over decades. When we met, he was immersed in an (ultimately successful) campaign to push Shell to abandon its plans to drill in the Arctic. John described climate change and biodiversity loss as the two biggest threats to our planet. He argues that ultimately the solutions are simple, and hinge primarily on reducing humanity's reliance on coal and livestock: two 'eminently achievable' projects for the years to come.

John is, however, concerned that demand for coal is still growing significantly, especially in emerging economies such as India, Indonesia and Vietnam – but also in the US, where President Trump has been providing state aid to the coal communities who voted for him in such strong numbers. John argues that to phase out coal is primarily a political challenge, rather than a technical or economic one: 'If governments decided today they wanted to transition out of coal, they could very easily do so. The case for the transition is there, and the benefits such as reducing air pollution clear.' It is the politics which need to be tackled: the coal sector is a powerful political force in most countries, and of course there must be a fair deal and a 'just transition' for the several million coal miners worldwide who stand to lose out from coal phase-out. Countries need to commit to retrain coal workers in other industries, including the burgeoning renewable industry.

John was furious that oil companies are still spending hundreds of billions on new oil exploration when we cannot burn the reserves to which we already have access. He is also incandescent about their 'criminally reckless' treatment of the Arctic, the 'earth's air conditioner'. He explained that for a long time, the Arctic had been considered a hostile environment and a wasteland, where it was impossible to fish or to extract oil; but that now, because of the changing climate, everybody was moving into this previously inaccessible area. 'The cycle is vicious and perverse,' he said; 'the Arctic is melting, leading to more fossil fuel extraction there, leading in turn to more melting and therefore to more climate change.' He went on to outline other alarming activities from these companies, such as BP's plans to expand into the Great Australian Bight, an extraordinarily pristine marine environment off the south-west coast of Australia and an important breeding ground for the Southern Right Whale. 'Oil companies have not changed their behaviour at all since the Paris Agreement,' he said. 'Investors still do not think governments are going to act, and so we have to hold governments to account.'

A different future is possible. If these same companies used their knowledge, skill and money to transition away from oil reserves and invest their know-how in a greener future, not only would they secure a market share in the energy system of the twenty-first century, but they would repay their societal debt by placing the world on the path to an inhabitable future. Whether or not they seek to be front-runners, or to dig their heels in, the changing economics of renewables, coupled with the technological transformation and growing levels of international climate action underway mean that the transformation of the energy sector is inevitable. The path to a new climate economy lies right before us, and it is ours for the taking.

But what about areas of the world where there is no national grid, and where people still have precarious if any access to energy? Surely these countries have the same right to develop on the back of fossil fuels as the industrialised world has? Rachel Kyte leads the UN Secretary-General's 'Sustainable Energy for All' initiative, where her brief is to 'illuminate a path by which the world can meet its 2030 Global Goal to achieve universal access to energy and energy efficiency'. The challenge is to harness the extraordinary growth in renewables – in which prices have fallen by 75 per cent – and triple the rate of energy efficiency so that the world can provide reliable, affordable, clean power to all. When we spoke, Rachel described a home in Nairobi, Dar es Salaam or Arusha, where a family with one small solar panel can have three super-efficient light bulbs, a radio, a small TV and a fan. As a result, they can charge a mobile phone, their children can read their homework with the light and a refrigerator can run. In a refugee camp, a solar installation built on top of a disused shipping container provides enough energy to recharge batteries and run a large-screen TV. In Kenya, Rachel described a factory run entirely by women making cookstoves which improve both the quality of air in people's homes and their quality of life.

To achieve Rachel's vision, the world needs more investment in smart grids, and better leadership from political leaders who understand that it is 'good politics (as well as being the right thing to do) for countries and societies to invest in the right, integrated energy systems for the future'. This is not an aspirational goal, but a necessity: some 1.2 billion people are still without access to energy, of which 620 million are in Africa, and the majority of the rest in Asia. There are also several million in Central and South America, with Haiti particularly poorly served, and many indigenous peoples without energy access.

Across the world, Rachel was heartened by many inspiring examples of the right things already happening: in Germany, communities run mini-grids alongside the main system; and in the US, there is much solar and wind off-grid, especially in areas where communities prize their independence from government. Even Mongolian nomads are using solar panels on the roofs of their yurts.

The world needs to understand that providing proper sources of clean, reliable power has valuable knock-on effects, including delivering better health care and broadband to isolated health clinics in both urban and rural areas. In almost all societies, there are also big emissions reductions to be gained through increased energy efficiency in industry, transport and buildings. Rachel explained that the whole world needed to become like Denmark on energy efficiency, and Denmark needed to scale up to a whole new level too. Everywhere, simple measures taken could lead to vastly more efficient systems, with substantial savings possible for companies, consumers and countries' balance sheets. In the UK, by way of one example, low-income families need support to improve the energy efficiency of their 'old and leaky' houses; refurbished housing stock typically uses new building codes with super-efficient appliances.

The future of the world's energy system is in Rachel's view 'decentralised, digitalised, decarbonised and democratised: therein is the reason for optimism. Communities can get power: nothing is more transformative.'

The world's efforts to bring about a renewable energy system would be greatly enhanced if countries, companies and jurisdictions put 'a big, fat price on carbon', in the memorable phrase of the Mexican economist and Secretary-General of the Organisation for Economic Cooperation and Development, Angel Gurría: a cost applied to the pollution generated by the global economy. A high-profile Commission on Carbon Pricing in 2017 concluded that a carbon price of at least \$40–80 per ton of CO_2 emitted by 2020, and of nearer \$50–100 per ton of CO_2 emitted by 2030, would be enough to drive a radical shift in the global economy consistent with meeting the Paris Agreement. We put prices on other negative aspects of the global economy, in the hope of encouraging better outcomes: why not do the same for this most vital of global challenges?

Encouragingly, at least 47 carbon-pricing regimes are already being implemented in 67 countries. The world needs to strengthen these regimes, make them stricter and encourage their adoption elsewhere. In addition, 1,400 companies are already pricing their own emissions, by using an internal carbon price to guide their long-term decision-making and business plans. This has led to companies behaving much more efficiently and ambitiously to reduce their emissions and make big savings in the process. Putting a price on carbon is one of the single biggest contributions we can make to achieving better climate policy. It offers a triple win: a positive contribution to the health of the environment and the public; an additional source of revenue to be spent on health, education and other spheres; and a spur to innovation and greater investment in cleaner, renewable technologies and companies.*

Linked to carbon pricing, there is an equally urgent need to reform the world's existing subsidies to fossil fuels. These currently equate to approximately $500 million every year, in the form of tax instruments, loans and grants to the incumbent oil, gas and coal sectors. Shelagh Whitley of the Overseas Development Institute is an expert on this issue, and she explained to me how much the world would stand to benefit both economically

---

* It was therefore heartening to see the economist William Nordhaus – a long-standing advocate of carbon pricing – win the Nobel prize for Economic Sciences 2018.

and environmentally from a redeployment of these subsidies, as a number of countries have already shown. India is one interesting example: with a clever advertising campaign, it has convinced wealthy and middle-class families to give up their subsidised gas cylinders, enabling subsidies to be directed solely at poorer communities. Subsidy reform in Indonesia had been linked to the provision of greater health-care access and better public transport and infrastructure to its people. One can only imagine what could be achieved with a portion of these savings invested in renewable energy research, energy efficiency and implementation. Indeed, if a fraction of the world's fossil fuel subsidies were redeployed to increase Mission Innovation's comparatively meagre budget, and thereby allow the renewables sector to compete on a more equal footing, our climate goals would be so much closer to being met. Ever since 2009, the G20 (Group of Twenty) countries have committed to eliminate fossil fuel subsidies, but progress is slow: the power of the incumbents is strong, and the political economy of reforming these subsidies not easy to overcome. But the case for reform is clearer by the day.

Another challenge is to make the world's infrastructure 'climate friendly' in the future: some $90 trillion will be invested in infrastructure in the next 15–20 years, the bulk of it in emerging economies. Countries can either choose to lock in their existing high carbon forms of infrastructure, or privilege the low carbon technologies of the future; they must not make the same mistakes industrialised countries have in the past. Fortunately, there are strong economic, human well-being and political arguments in favour of better, greener, more low carbon infrastructure.

As Andrew Steer, president of the World Resources Institute, told me, the people of Mexico City have a keen interest in ensuring that the future transport systems of their city reduce pollution and avoid gridlock, not least because it delivers people from home to work and back in good time, meaning parents can spend more time with their children, and put them to bed at night. That such public transport systems are so manifestly desirable from these perspectives, almost irrespective of their climate benefits, makes them – or at least should make them – a winning proposition in all countries.

And so it is that the vision of a global economy based entirely on low carbon, renewable energy is a fundamentally achievable one: a world in which energy-efficient industries and buildings are the norm; in which everyone has access to sufficient, reliable clean energy; in which clean public transport systems capable of transporting millions of people without significant emissions have become widespread; and in which whole factories and industries are also run safely and efficiently on renewable energy.

We are making strong steps, emboldened by amazing technological advances, but we need to move much faster to win this race against time. Johan Rockström, the distinguished Swedish scientist who has been at the heart of recent scientific thinking about planetary boundaries, biodiversity and climate change, argued, in the weeks leading up to Oli's birth, that the world needed to halve its emissions every decade to meet the Paris Agreement. He proposed this ambition be called a 'carbon law': a variation on Moore's Law, in which the power of computer processors is held to double every two years. In Rockström's model, fossil fuel emissions should peak by 2020 at the latest and fall to zero by 2050. The amount of renewables in the global energy system will therefore need to be doubled every 5–7 years. The Rockström model also requires that the world achieve net zero emissions from land use – agriculture and deforestation – by 2050 (to which I will return in the next two chapters). Additionally, the world will need to invent technologies to remove 5 gigatons of carbon dioxide from the atmosphere per year by 2050. In Rockström's words, 'It's way more than adding solar or wind. It's rapid decarbonisation, plus a revolution in food production, plus a sustainability revolution, *plus* a massive engineering scale-up for carbon removal.' To put a plan like this into action is the abiding challenge of our times.

For many people around the world, meanwhile, climate change is already leading to daily, measurable impacts on their lives. Seasons are changing, working in agriculture is getting harder, and heatwaves in cities are leading to premature deaths, particularly among vulnerable and elderly populations. In areas of conflict and humanitarian crisis, climate change has played a 'multiplier' effect in rendering complex, multi-faceted situations more volatile

– and this trend is set to continue if climate change continues unabated.

One emblematic case is Syria, where an intense drought from 2006 to 2011 led to failing crops and the migration of 1.5 million Syrians to urban areas, adding fuel to the fire of underlying social and political conditions which led to the devastating civil war over recent years. Climate change did not cause the Syrian civil war: far from it. But it did play its part in exacerbating the underlying conditions which led to war. There are other areas, across the Middle East, North and sub-Saharan Africa, where communities already living in great need have found themselves in even more precarious or tense situations as a result of droughts, sporadic rains and excessive heat. These tensions are only likely to increase.

The extent of the threat posed by climate change has led the UN Secretary-General, António Guterres, to make adaptation to climate change, resilience and disaster preparedness a key priority for his administration. The UN is working hard to ensure countries invest more in these areas as a means of preventing war and humanitarian crisis. Given that climate change, water scarcity, desertification and environmental degradation are already making life very much harder for millions of people around the world, there is a moral imperative to invest in a far-sighted way to reduce these impacts.

In late 2006, I spent several weeks in Yemen where Emma Nicholson, then a Member of the European Parliament Foreign Affairs Committee as well as director of the humanitarian organisation AMAR Foundation, for which I worked at the time, had been asked to lead the EU's Election Observation Mission.

We travelled right across Yemen, from north to south, meeting politicians from the steely President Ali Abdullah Saleh (killed by Houthi rebels in December 2017) to the democratic opposition, sitting on the floor during long meetings of Yemen's parliamentarians chewing the narcotic drug *khat*, inspecting polling booths and election stations, and talking to electoral officials. Across the land, even then, there was much talk of a looming environmental crisis, with freshwater supplies being drained – in part to grow fields of water-intensive *khat* – and climate patterns becoming ever more varied, with food shortages already being felt. Even in 2006, Sana'a vied to be the first of the world's cities to run out of water (although in February 2018, this happened in Cape Town first), with talk of the city needing to be relocated elsewhere.

More than a decade on from that visit, Yemen is now undergoing a terrible famine, made worse by Saudi Arabia's war with the country, which has led to aerial bombardment and blockades of key Yemeni ports (with the UK and others continuing to sell arms to the Saudis). I recall vividly how one night, after a day of election monitoring, the Irish expert on Yemen's electoral arithmetic, Michael McNamara, and I hitched a lift with our EU car (and minor military entourage) to the old fourteenth-century town in downtown Sana'a, where we sat in the square at a long table with dozens of Yemeni men, wearing their traditional cloth dress and ceremonial swords. Here, young boys serving as waiters brought us delicious lamb kebabs from coal fires, accompanied with fresh hot bread and tea; all under a moonlit sky and surrounded as far as the eye could see by the wonderful, gingerbread-coloured, turreted sandstone buildings which had made Sana'a famous – many of them now crushed in all the bombings and warmongering of recent months and years. When I think of the impact of climate change in the world, my first thoughts are with Yemen; but every country has its story to tell.

The world therefore needs a renewable energy transformation as fast as we can possibly manage, while we must also find new technologies with which to remove carbon from the atmosphere; in no scenario is our current reliance on fossil fuels sustainable. The longer we wait, the worse the damage to the planet and to us becomes. To create a world in which people no longer have to migrate as a result of a deteriorating environment and climate heading out of control, we also need more investment in enhancing the underlying ecological resilience of communities currently living in such vulnerable environments. The next three chapters describe how.

While it will be impossible to forestall some of climate change's most intense and inexorable impacts, and we will need to adapt to the changes which are already upon us, there is still so much we can do to prevent the worst. In the run-up to the Paris talks, the journalist Thomas Friedman documented how drought and water scarcity in a village in rural Senegal left a father of four no choice but to leave his wife and children behind, to cross Libya and find his way at considerable cost and risk on to a precarious boat to Europe, in the hope of finding refuge and eventually work to send money home to his family. The level of desperation and absence of alternatives that lead someone to give up their home and leave their family for such an uncertain future are striking. This kind of migration is sure to happen at a much greater scale if the climate continues to change in the ways the science indicates is likely. As I write, there is drought and famine in Southern Sudan as well as in large parts of Ethiopia, Nigeria, Somalia and Yemen, driven predominantly by conflict and politics, but undeniably exacerbated by the underlying climatic and natural resource constraints increasingly faced by those countries. We must move fast to bring about the energy transformation the world urgently needs to arrest these changes and ensure a better future for us all.

PATH TWO

Forests

I have spent the past ten years or so of my life, in various guises, working to encourage governments, companies, philanthropists and communities to protect and restore the world's tropical rainforests. If I had a magic wand, and could achieve just one of the many ideas put forward in this book, it would be to save those remaining tropical rainforests. As I write these words, the organisation for which I work – the World Resources Institute (WRI) – is involved with partners in an effort to support and encourage the government of Indonesia to protect the vital tropical forests of Papua and West Papua. WRI has also published the most recent global deforestation figures: 15.8 million hectares of tropical rainforest were lost in 2017. The plight of forests and rainforests worldwide is a cause which takes you by the scruff of the neck and will not let you go.

The rainforest I know best is in Colombia, a country where I lived and worked for a number of years in the President's office as an environmental adviser. While in Bogotá, I chaired two round tables on the future of the

Colombian Amazon and the Biogeographical Chocó, a precious area of the Colombian Pacific Coast. Both of these regions are home to truly astonishing rainforests, and I had the privilege to travel in both at length. I then worked for six years for an organisation established by the Prince of Wales to promote forest conservation, the International Sustainability Unit (ISU), and this work took me to Brazil, Indonesia and the forests of West and sub-Saharan Africa.

Colombia set the world a striking example of how to protect forests and indigenous peoples in the late 1980s. The then President Virgilio Barco declared that at least 50 per cent of the Colombian Amazon should be formally and legally protected as indigenous reserves. The area – not far off the size of the United Kingdom – has benefited from comparatively high rates of conservation and forest cover since then, although there is now more development pressure in these areas in 'post-conflict' Colombia, with inevitable investments being made into roads, oil, mining and agricultural development.

My friend and mentor Martin von Hildebrand, the Colombian anthropologist, played a vital role in convincing President Barco to make this decision. The two would meet over lunch or for a whisky at the end of the President's long day, late in the evening, and Barco would ask after the well-being of Colombia's indigenous people with sincere concern. Martin would tell him stories about his recent visits to these communities, where he would often spend months at a time, and gradually convinced the President that what the people really needed above all was proper and legally binding recognition of their land. And so it was that Barco made the commitment to reserve large areas of forest land in perpetuity for the indigenous peoples. The news gained international press coverage, including in the UK, where the Prince of Wales lauded the Colombian president's vision.

The establishment of the indigenous reserves in Colombia mirrored and encouraged the development of similar legal rights for indigenous and forest peoples to their lands elsewhere in the Amazon Basin, including in Brazil, Peru, Ecuador and Guyana. While indigenous peoples' rights are often not honoured on the ground, and there is a constant need to hold governments and others to account for their enforcement, there is strong evidence that indigenous reserves (and protected areas) lead to significantly greater levels of forest cover over time. To demonstrate the strength of the overlap, simply transpose two maps of the Amazon Basin: the first, a map of deforestation across the region; the second, a map of indigenous reserves and protected areas. With few exceptions, you will see that most deforestation is occurring in areas which are designated neither as indigenous reserves nor as protected areas.

In 2011, when still based in Bogotá, I supported the Colombian government's efforts to expand Chiribiquete National Park, an area of outstanding natural beauty and cultural significance in the heart of the Colombian Amazon, from an area of what was previously 1.2 million to over 4 million hectares. The enlarged park – if it can be enforced on the ground, always a challenge in areas where the presence of the state and the park authorities are weak – guards the heart of the Amazon from agricultural development. The national park was one part of a much broader set of policies intended to protect the Colombian Amazon, encompassing alternative livelihoods for local communities, as well as an effort to establish a much stronger and more effective state presence in areas which had historically had little or no such presence.

I once had the great joy and privilege of going on a week's odyssey with Martin and the celebrated Colombian writer Héctor Abad Faciolince to the indigenous territories. We flew to the heart of the Amazon on a small four-seater plane, its dappled shadow memorably etched, in a scene otherwise reminiscent of *The English Patient*, on the verdant green canopy of the forest, and then travelled by boat and on foot for the rest of the way. It was a wonderful experience. We ate with the communities and spent long hours talking with them in the *maloka*, the community hall, late into the night, seeking to understand their vision of the world and of their environment; sleeping in hammocks; drinking fermented corn beer; chewing *mambe*, the mashed-up coca leaf, which dries the mouth, and sometimes makes speech difficult, but which also seems to make one's thoughts more lucid and cogent; going hunting in the river with the young men (to my enduring shame, they caught a young crocodile which we ate for breakfast in a salty soup the following morning); washing in the river; and so on.

The forest was in remarkably good condition, with only minimal amounts of wood being harvested to help with building schools, *malokas* and fishing boats. There were however some signs that the forest's fauna had at times been over-hunted, and the indigenous people were not allowing populations of some species to recover as, the elders told us, their forebears had once learned to. The rivers were also being overfished, in a similar fashion as I had witnessed with the poor little crocodile. But there were plans afoot to reset the balance and to encourage the communities to adopt best practice again.

The more existential threats to the communities' existence in the long run, however, came from further downstream: a Canadian gold mine was pressuring villagers down the river to allow the company to operate in the forest, on the promise of jobs and a share of the wealth. There was a real risk that this mining would contaminate the river and damage the livelihoods of the groups further into the forest. We also wondered whether the communities would continue to live in isolation, given that they were several days' journey by river from the nearest medical centre. These were some of the questions we discussed amid the dense forest, black rivers and starry nights of the Colombian Amazon. After a week, we travelled back to the village where we had landed, sat in the village school and waited for our small plane to pick us up.

Since that journey, Martin tells me that the cultural integrity and commitment of the villagers to their way of life stays strong, and that the majority are continuing to withstand the pressures of those that wish to exploit the forest. But these do not go away, and indeed are likely to increase.

There are other examples of success in protecting and restoring tropical forests. Costa Rica, having deforested significantly from the 1970s to the 1990s, has managed to increase its forest cover by 20 per cent over the past two decades, establishing a nationwide ecotourism industry based on the health of its forests. When the bottom fell out of the livestock market in the 1990s due to its currency deflation, Costa Rica re-evaluated its options, and the transition began. The country now has some 60 per cent forest cover, as well as a strong agriculture sector and a diversified economy. Ecotourism generates a significant portion of the country's revenue, such that Costa Rica is now an exemplar – including in David Attenborough's estimation – of how a country can develop in a way which protects, rather than damages, its natural heritage.

For much of the past decade, Brazil was also held to be a significant success story, although there have been some worryingly negative developments in the past few years which have put this achievement seriously at risk, particularly with Jair Bolsonaro's election in 2018. Previous governments tackled deforestation, which by 2014 led to an 80 per cent reduction from 2006 levels – and while the ongoing deforestation still represents a huge area (at least four times the size of São Paolo is still lost every year), this reduction was a globally significant achievement. Three factors played a key role.

The first was the commitment of the former president, Luiz Inácio Lula da Silva, now mired in a corruption scandal, a popular figure who positioned Brazil as a leader on climate change and sustainable development in the run-up to the Copenhagen Climate Summit in 2009. Lula – egged on by his charismatic environment minister, Marina Silva, in my view one of the world's most inspiring environmentalists – instructed the Brazilian authorities to police deforestation more actively, both on the ground and using state-of-the-art satellite technology, fining municipalities still in breach of their environmental commitments, and rewarding those which had achieved reductions.

The second factor was an engaged private sector – soya and beef traders, McDonald's and others – which had been propelled into action as a direct result of some highly effective and targeted campaigns from Greenpeace and others in Brazil. The companies placed a moratorium, which still stands, on illegally sourced beef: if the traceability of the beef was not assured to be from good sources, it would not find a market. A similar moratorium was placed on deforestation-free soy, which is currently under review for renewal.

The third factor was that farmers who had been historically responsible for deforestation became beneficiaries of increased state support to prevent it. With these funds, they were able to improve their agricultural practices and increase their yields, as well as receive technical support with managing their land and registering their farms on an online monitoring database.

Brazil's success in reducing deforestation in the Amazon has been undone by the country inadvertently increasing agricultural pressure on another vital ecosystem or biome in the country, the Cerrado. Some 80 per cent of the Cerrado's original vegetation, forest cover and biodiversity has been lost as a result of agricultural expansion in this region over recent decades. Acid soils have been rendered productive through the use of massive amounts of limestone and fertiliser to produce maize and soya for the global market. The impacts on the ecosystem have been highly detrimental, with many native species lost and little short of an ecological calamity wrought on what was once an intricate patchwork of a landscape. A number of national and international non-governmental organisations (NGOs) have recently issued an urgent plea, in the 'Cerrado Manifesto', calling for great care and attention to be paid to protecting the fragments which remain.

Brazil is also home to some important remaining rainforest on its Atlantic coast. The Mata Atlantica may be a fraction of its former self: 95 per cent of what was once the Atlantic rainforest has been deforested since the arrival of the Portuguese in the sixteenth century. But the Brazilian government has made ambitious commitments to protect and restore large swathes of what remains, in a patchwork approach involving coffee, acai (the popular berry, available in smoothies on every street corner in Rio), ecotourism and timber production. Over recent decades, tens of thousands of hectares have been committed to diverse forms of ecological restoration. Nature, as so often, given half a chance, recovers fast: there is always hope. At the political level, one can only hope that the new Brazilian government – led by the nationalistic right-winger Bolsonaro – will find a sense of environmental commitment and responsibility, and withstand the pressures of a rampant agricultural and rural lobby, for its and the world's benefit. Marina Silva is continuing to argue for such an outcome, but her voice is, alas, a fairly lone one in the Brazilian polity at the moment. It is no exaggeration to say that the future of the world hangs on how Brazil behaves towards its Amazon Rainforest.

In the Gola Rainforest spanning Liberia and Sierra Leone, a number of local NGOs, supported by Birdlife International and the Royal Society for the Protection of Birds, have worked for decades in partnership with local communities and national governments to protect and restore some 350,000 hectares of still highly biodiverse rainforest, set within a landscape of cocoa plantations and agroforestry. With patient and respectful work with local people, coupled with a real attempt to ensure that these countries' presidents and ministers feel proud of and responsible for these areas, work in the Gola has shown that it is possible to build a lasting constituency for forest protection in countries and forest regions.

In addition to protecting remaining areas of standing forest, replanting trees and reforesting is possible across Africa (as well as Central and South America and Southeast Asia). In Kenya, for many decades, the inspiring forest leader and activist, Wangari Maathai, led the Green Belt Movement, which has reforested large areas of the country, and her daughter Wanjira continues to fly the flag for the cause today. The Prince of Wales planted a tree in Wangari's memory at Kew, at a moving ceremony on a cold winter's day in 2013. In his eulogy, the Prince said: 'She understood the link between poverty and the natural environment. We are faced with so many massive challenges that at times it is utterly overwhelming. We have the responsibility to protect the rights of all generations of all species who cannot speak for themselves.' Similar commitments and social movements are under way in Ethiopia, Niger, Mali, Malawi and elsewhere across Africa.

The 'Great Green Wall of Africa' is an African-led initiative to restore the semi-arid Sahel – from Dakar on the west coast to Djibouti in the east – into a fertile land, capable of withstanding the pressures of a changing climate and an encroaching Sahara Desert. This is easy to say and difficult to do: many of the countries of the Great Green Wall face daunting poverty and governance challenges. The arid landscapes are themselves difficult to restore, but real and heartening progress is being made in each of the countries which has signed up to the initiative.

Sabah, in Malaysian Borneo, after decades of deforestation, has an ambitious plan to protect and restore large areas of the forest by showing that it is possible to source commodities in a sustainable way. The plan could lead to the long-term protection of the beautiful landscape of the Danam Valley, where a charismatic British horticulturalist, Dr Glen Reynolds, has devoted years of his life to the protection of abundant Bornean gibbons, sambar deer, pygmy elephants, orangutans, civets, leopard cats, flying foxes and bearded pigs. Although there has been a lot of hunting and 'defaunation' of Borneo's rainforests, Danam Valley has remained largely ecologically intact. Jurisdictions such as Sabah are seeking to encourage supermarkets and traders to commit to sourcing their coffee or palm oil from areas where land-use planning and proper enforcement have put a stop to deforestation and encouraged the take-up of better agricultural practices across an entire region. Unilever has recently committed to sourcing sustainably produced commodities from Sabah.

The Great Bear Rainforest in Canada is the world's largest expanse of coastal temperate rainforest, and in 2016, a new agreement to protect large swathes of this 21-million-acre wilderness – the so-called 'Amazon of the North' – was made. This area is home to 1,000-year-old cedar trees, waterfalls, moss-covered mountains, fjords and dark waters. It is also home to indigenous First Nation communities who continue to call the forest home, as well as abundant wildlife such as grizzly bears, Sitka deer, grey wolves, sea otters, salmon, orca and humpback whales off the coast, and the much-venerated spirit or Kermode bear, emblematic of the region. The 2016 agreement now puts aside at least 85 per cent of the remaining forest in perpetuity, exempt from any logging or exploitation (although there continue to be worries that Prime Minister Trudeau will overturn his original opposition to enabling an oil pipeline and tankers to operate in some areas of the forest).

The battle against climate change can only be won if we manage to maintain a world of abundant, healthy, resilient forests. Forests, both tropical and boreal, play a vital role in regulating the climate and are vital to people's well-being, to biodiversity and to a flourishing agricultural system.

As the WRI numbers show, on the order of 16 million hectares of tropical forest are still either cut down completely, or substantially degraded, every year. There are many causes, including infrastructure development (roads, mines, cities), logging for timber, forest fires and the continuing expansion of the world's agricultural frontier as we scramble to meet growing global demand for commodities. Indeed, agricultural development is currently responsible for at least half the deforestation occurring in tropical zones, driven by the world's seemingly insatiable demand for palm oil, beef, soya, cocoa, sugar, coffee, rubber, pulp and paper.

In temperate and boreal zones, forest fires, disease and

excessive logging are also causing grave threats to forests across the US, Canada, Russia, Europe, Chile and elsewhere. Most recently, California, Greece and Portugal have experienced devastating and horrifying forest fires, with apocalyptic scenes of people fleeing or trapped in villages and houses surrounded by flames. The fires in California in late 2017, and again as I write these words in August 2018, have been the largest in the state's history. Spurred on by unusually intense winds, and made more destructive as a result of population growth and urban expansion, they have torn through many of the region's most iconic vineyards and landscapes.

The continuing loss of the world's tropical forests is not simply a tragedy in terms of our efforts to maintain a stable climate: they are also havens of biodiversity, home to many millions of species not yet known to man, including, quite possibly, cures in plants to some of humanity's most pressing ailments. We have barely any sense of the diversity and potential treasure trove that forests contain. Tropical forests are also fundamental to the water cycle, providing the rainfall on which agriculture depends through the 'rivers in the sky' that they generate. They are also vital to the survival of some of the world's oldest and most diverse cultures, and indigenous and forest peoples. Even if the climate were not such a pressing concern, it is still vital to protect and restore forests across the world. Despite the scale of the losses, there are dozens of inspiring examples of countries, communities and companies that have taken action to protect and restore their forests, with often startling results. While the overall trend is bleak, we must not lose heart, because we have learned how to stem the tide of deforestation in many areas, and to ensure that local people benefit from the protection of this vital asset. As the Chinese proverb puts it: 'The best time to plant a tree was twenty years ago. The second best time is now.'

To protect and restore the world's remaining tropical forests, we need to start by giving and securing indigenous peoples' rights to their ancestral land, as I learnt in the case of Colombia, and by upholding these rights against a multitude of pressures. More protected areas need to be established and better implemented on the ground: together, these two measures are our best hope for protecting what remains.

At the forest frontier, where the need for more land for crops threatens to encroach further into intact forest, we need urgently to find ways to improve agricultural production, so farmers on the periphery are less driven to cut down more forest for new land. In Indonesia, for instance, various agroforestry schemes provide alternative livelihoods in the areas adjoining the remaining areas of intact forest, which generates real income for local people while safeguarding the forest. The world also needs to put pressure on the large agricultural commodity companies, which source the foods we all eat, to eliminate deforestation from their supply chains. There is no excuse, at a time of such unparalleled technology to map and trace commodity supply chains, for companies not to take responsibility for the social and environmental conditions in the landscapes in which they operate. By the same token, banks and investors should divest from companies with a negative forest footprint, and invest in funds which protect the forest. Companies and banks in emerging economies also need to apply this same level of concern and anti-deforestation criteria to their operations. We risk, otherwise, a two-tier market in which Western-facing brands and consumers source sustainably produced palm oil to conscientious buyers of oatcakes in Sainsbury's, while India, Indonesia and China source much greater amounts of palm oil for domestic cooking with no such regard for the sustainability of its production.

The good news is that forests grow back remarkably quickly, given half a chance. They can and must play a vital role in the world we wish to see. Societies, communities, individuals, political leaders, faith groups and companies should rally behind the better conservation of the world's natural ecosystems – our forests above all, but also our peatlands, wetlands and savannahs – for the benefit of the world today and in the future. This is an abiding challenge for our times, and one which the natural world, with our help, can rise to meet. But we have no time to lose.

Soil

The protection and restoration of the world's vital soils is our third pressing path to a hopeful future. The great American writer Wendell Berry put it well, when he wrote: 'The soil is the great connector of lives, the source and destination of all. It is the healer and restorer and resurrector, by which disease passes into health, age into youth, death into life. Without proper care for it we can have no community, because without proper care for it we can have no life.' As a result of the impact of agriculture and human intervention, some 2 billion hectares of the world's land surface are now degraded. With the right set of policies and interventions, a great deal of this land can be rescued and restored to good health. Given half a chance, nature is wonderfully capable of recovery and restoration. Three examples from different parts of the world show what is possible.

The first is in the Loess Plateau (Huangtu Plateau), in north-west China, which is home to over 50 million people. Over two decades, this dry and windswept area that had been characterised by uncontrolled grazing, subsistence farming, unsustainable harvesting of fuel-wood and precarious cultivation of crops on slopes was restored to an area of abundant grasslands, trees and shrubs. Juergen Voegele, a German agronomist from the World Bank, believes the success of the project came from listening to local people, understanding their perspectives and then enlisting their support for the restoration which followed. For the duration of the project, Juergen travelled from village to village, winning support for the restorative approach to farming advocated by the local authorities and the Bank. To this day, the restoration of the Loess Plateau gives him great hope that similar feats could be achieved around the world.

The restoration of the plateau led to a dramatic reduction in soil erosion and the sedimentation of the region's waterways, saving on the order of 100 million tons of soil every year. A network of small dams ensured better sediment control and reduced the risk of flooding. The project focused on finding more efficient forms of crop and livestock production. Whereas before frequent droughts had caused the crops on the slopes to fail, the restoration work established terraces which made the crops much more resilient. Over two decades, the whole area was transformed, with lasting implications for the people of the region, and the approach is now being adopted at a national scale.

Another powerful example of soil restoration occurred in Ethiopia. In 1984–5, the country suffered a terrible famine which led to the loss of many thousands of lives and an international humanitarian campaign which people remember to this day. In 2007, the Ethiopian government started a land restoration campaign to plant 60 million trees, a commitment which in 2014 was raised to 22 million hectares: an area more than one-sixth of the entire country. By 2036, the national government aims to recover and rehabilitate up to 33 million hectares of degraded land, and to ensure that the country's agricultural production area remains stable without further incursion into the country's biodiverse forests, wetlands and other vital ecosystems. In a number of areas of Ethiopia, communities which have restored land have been shown to be more resilient to drought and more able to produce the food they need to sustain themselves. Restoration activities have also created more jobs, led to the replenishment of groundwater sources and fostered a stronger sense of attachment to the land.

One area in particular – the Tigray region of northern Ethiopia, in a remarkable effort led by community leader Aba Hawi – has shown the extent of what is possible in restoring the earth. Through concerted effort, and with inspired leadership, the Abreha we Atsbeha community has brought back water and vibrant colour to what was once a desolate landscape marked by drought and human suffering. The establishment of soil and water conservation structures on the hillsides to counter soil erosion, coupled with the planting of trees and shrubs, ensures that every drop of water that falls during the region's two-month-long rainy season is captured in the soil. Communities earn precious income from the honey and the fruit trees they produce.

The *Faidherbia albida* tree has been instrumental to the restoration achieved: this species sheds its leaves in the wet season and regains them in the dry season, when fodder is scarce, providing shade and nourishment to the soil. The African juniper, the sand olive and the avocado tree have also been used to great effect in the restoration. Bench terraces, with deep trenches dug by community volunteers, have improved the recharging of ground-water, with waterfalls and natural springs appearing again. Sections of the land have also been closed off to allow natural regeneration to occur, ensuring that the free grazing of livestock does not come at the expense of the young trees and saplings. In summary, the restoration of these iconic areas in Ethiopia has been an astonishing achievement, and an enduring inspiration for the rest of the world to follow: an area associated with human tragedy a few decades ago is now home to viable rural communities drawing a sustainable living from the land, and seeking to remain there for the future.

Similar success has been achieved in Niger: one farmer's successful restoration of his land led to a country-wide movement which has transformed an area of over 5 million hectares. This approach involves fencing off tree saplings to prevent them from being trampled on; the creation of vegetable gardens to improve the food security and diet of the farmers; digging deeper plant pits to enable water to collect; better water-harvesting techniques; and micro-dosing of mineral fertiliser, rather than the previously more wasteful approach to fertiliser use. Employed together, this approach to 'farmer-managed natural regeneration' has inspired a nation to think differently about its rural environment. The farmer's techniques spread from village to village and are now in operation in large areas of the country. As the health of the land improved, the water level in the aquifers in one area rose by as much as 14 metres. Women who walked half a day to fetch water now walk half an hour to the closest well.

In some areas, cattle can play a role in regenerating ecosystems. One celebrated example has been described by the Zimbabwean farmer Allan Savory, who discovered that if cattle could replicate the traditional grazing patterns of wild animals, in which they move en masse from one area to another, previously degraded soils can be regenerated and thus store more carbon. This insight originated from Savory's observation of the wide plains of the Serengeti, where wild animals such as wildebeest live in large herds, often made up of several hundred thousand animals. The animals stay close together, as a result of the 'herd mentality' and to protect themselves against predators: in southern Kenya, my wife Davina and I saw how those wildebeest which stray far from the herd, or are left behind by it, get picked off by lions, cheetahs and crocodiles. In pursuit of fresh grazing land, the wildebeest are constantly on the move: they settle in one area for a day or two, trample all over it and leave their dung, and then move on elsewhere, not to return for months. These natural migratory patterns are at variance with the bulk of today's cattle management systems, where cows are typically confined to one fixed area, where they eat all the buds and shoots of a field and have a negative impact on the ecological health of the landscape. In the Savory scheme, cattle are brought together in a single herd, and moved from paddock to paddock in a planned, 'rotational' grazing system. The cattle trample the soil and have an intense impact for a few days, but they then move on: the fields recover and regain life and health, sequestering significant amounts of carbon in the process. The system is now being used widely and with often remarkable impacts in Australia, Chile, Argentina, the US, Mexico, the UK and elsewhere. There is still a place for cattle in our world, but under these kinds of systems, in which the soil, the trees and the wider landscape can also flourish.

Sitting on the grass in Green Park, Satish Kumar, the Indian environmentalist who has campaigned on environmental and peace issues since the 1950s, explained to me: 'Soil is life; soil is everything; without soil, there is nothing. Soil, soul and society' are the key to saving the planet.

We have gravely mismanaged the world's soils, an estimated third of which are now moderately to highly degraded due to nutrient depletion, erosion, acidification, salinisation, chemical pollution and compaction. Each year, the world loses some 24 billion tons of fertile soil as a result of the way we farm and manage our environment.

The earth cannot afford to continue like this. Instead, we need to find new ways of protecting and restoring the soil through better farming techniques which safeguard the soil, rather than deplete it. The farming systems of the future will be capable of achieving an impressive balancing act: namely, to meet humanity's needs for healthy and nutritious food, while providing better lives for the world's 500 million smallholder farmers, all the while ensuring farming practices which are good for the climate and kind to biodiversity.

Fortunately, there is plentiful evidence from around the world of better agricultural systems generating the food we need, while improving the livelihoods and well-being of farmers, and protecting vital ecosystems, such as forests, wetlands and mangroves. But the challenge is massive. Today, as a result of fertiliser use and run-off, soil erosion, creeping desertification and tropical forests being cleared, global agriculture and land use contributes more emissions to the atmosphere than all the world's industry or transport – approximately a third of the world's total greenhouse gas emissions. A veritable transformation is required, in short order, to deliver an agricultural system based on the health of the soil and its regeneration.

I decided to begin my explorations in this area by seeking to understand the perspectives of an organic farmer, and so turned to Lord Peter Melchett, Policy Director of the Soil Association, a long-standing and much-loved advocate of organic agriculture as the best and most sustainable means of feeding the world.* Peter was once a Labour politician in the Harold Wilson government and subsequently campaigned for Greenpeace, becoming famous for uprooting genetically modified (GM) maize crops under trial in the late 1990s. Peter invited me to his organic farm in Norfolk during the summer of 2016 for a day's walk through his flower-lined fields. At the time, he and I were involved in an effort to support the French government's new initiative to push for greater global investment and action to improve soil health, and so a lesson in healthy soil felt particularly welcome and relevant when we met.

The French had launched their initiative in the months leading up to the Paris talks. Their argument was that if the world were to increase the quotient of carbon (or soil organic matter) contained in its soils by a factor of 4 parts per 1,000, this would lead to major greenhouse-gas emissions reductions, as well as to more sustainable and resilient forms of agriculture, beneficial both to farmers and to global food security. The scale of the prize in climate terms was massive: as much as a quarter of global emissions could be mitigated in this way (a claim about which many scientists are uneasy). The phrase '4 per 1,000' was the brainchild of the former French Minister of Agriculture Stéphane Le Foll. He is a charming and charismatic man, a socialist with big farmer's hands whose meeting table in his elegant

---

* Very sadly, Peter died on 29 August 2018, in the latter stages of editing the book. I honour his memory and his legacy here.

eighteenth-century office in Paris had a large ashtray on it, the smell of cigarette smoke very much part of the atmosphere. On the day we met, he had just concluded detailed and fraught negotiations with France's dairy farmers: there was a noisy press conference in the seventeenth-century courtyard next to his office right before our meeting.

While in office, Le Foll had become a passionate advocate for a global focus on boosting soil carbon, and had recently been travelling in the US and West Africa, as well as across France, to make his argument. In the US, he had been to Ohio – an epicentre of global soil research, and home to the doyen of soil scientists, Professor Rattan Lal – and visited the farm of David Brandt, one of the pioneering American 'carbon farmers', a form of agriculture which explicitly focuses on putting the carbon back into the soil. In West Africa, he had seen how poor farmers could improve their livelihoods in Senegal, Mali and Burkina Faso through better farming practices focused on greater care for the soil, and encouraged the French Development Agency to support these efforts further. And in southern France, as a photograph on his mantelpiece attested, he had seen how farms and vineyards had improved the carbon and nitrogen content of their soils by planting trees throughout their fields as well as leguminous, nitrogen-fixing crops.

Back in the UK, Peter Melchett had learnt of the French initiative with some excitement and was attempting to persuade the British government to sign up. And so it was that I found myself in Courtyard Farm, Peter's family farm in the heart of Norfolk, where he is proud to have built up the carbon content of his farm's soils over fifteen years, with the records to prove it. In Peter's office, we pored over historical maps of the landscape, photographs and records of the farm which dated back to the seventeenth century, as well as his more recent data and charts. By careful application of organic and ecological agricultural approaches, soil organic matter had gone up by a factor of two and a half times since he converted the farm to a fully organic system, with the trend set to continue.

On a long walk through the fields, we came across Peter's son driving a tractor; father and son compared notes on how the fields were faring in light of the recent weather. Peter explained how he avoids artificial pesticides or fertilisers, and provides a habitat for wildlife by protecting riparian borders and cultivating wildflower meadows, clover fields and areas of natural beauty to benefit insects and native plants which in turn attract mammals and birds.

Peter's clover fields – 'usually now a mixture which is mainly red clover, with some white clover, and other nitrogen-fixing legumes like trefoil and vetch' – are an essential means of putting nutrients back into the soil. Energy from the sun allows the clovers (as well as the peas and beans) to fix nitrogen in their roots, vital for crop-growing. After growing under a wheat crop for one year – really just at ground-level, and nowhere near the height of wheat – the clover remains undisturbed for another two years to maximise the nitrogen-fixing. The clover fields are then cut for silage, to feed the cattle in winter, or to be grazed by young cattle in the summer; the pigs are also put on the fields during the final year. Over the winter, the farm grows a cover crop, usually mustard or vetch, to prevent any soil erosion and to maintain the nutrients for the next crop to be planted in the spring. The farm grows wheat, barley, beans, peas and red clover in a six-year rotation. The wheat, barley, peas and beans are grown for use as seed by other organic farmers, which means the fields have to be clean and the grain of high quality. The peas and beans are used primarily for animal feed (thereby substituting soya grown in the Brazilian Cerrado); the wheat for animal feed or to be milled for biscuits and bread; and the barley for beer malting or for animal feed. 'Rotations will vary with the strength of the soil, farmer preferences, weather and other factors,' Peter explained. His fields of organic wheat and barley are selling at a favourable price compared to neighbouring non-organic farms, due in no small measure to the ongoing market for organic bread, both in Norfolk and further beyond.

The farm adheres to the Soil Association's most rigorous organic standards, and employs more people than the intensive farms in the area (some of which employ fewer than one farm worker per 1,000 acres, while Courtyard Farm had a full-time manager and regular part-time workers on its 890 acres, in addition to three full-time workers, plus the need for additional temporary help with the pigs).

The pigs were not there, alas, when I visited, but Peter

was proud of the arrangement he had struck up with the Organic Pig Company which produces organic pork for Waitrose. Some 750 sows and their young live on approximately 90 acres of the farm's clover fields, and are moved to fresh fields after one year. The sows are allowed to make a straw nest to give birth, and the pigs and piglets are fed free-range, organic and non-GM diets, as well as being left to root around in the rich soil for insects and roots. Wintering birds – including jackdaws, starlings, gulls and wagtails – benefit from the extra food provided by the pigs, and Peter noted too that the pig manure adds phosphate and potassium to the crops, just as the clover fields fix additional nitrogen to the soil. His pigs and cows grow more slowly, eat more natural diets, spend more time with their mothers, move around more freely and live in more settled family groups to minimise stress. The farm's commitment to animal welfare is strong and central to its farming.

At the end of our walk, Peter showed me the three small 5-kilowatt wind turbines installed on the central site of Courtyard Farm, sufficient to run all the farm's buildings and operations, and I thanked him for a wonderful lesson in the joys and benefits of a system which, in Peter's devoutly held view, is sufficient both to feed the planet and to meet our climate and biodiversity goals at the same time.

Some experts question whether organic farming systems such as Peter's, however desirable and bucolic they may be, are the only answer to the world's food security and agricultural challenges, especially in developing countries. To seek a further set of perspectives, I then sought out Professor Sir Gordon Conway, who has worked for over forty years on agriculture in developing countries, for an interview over a cup of tea on a cold winter's day at Imperial College just before Christmas in December 2016. While Gordon greatly respects the principles and practices of organic farming, he argued that a middle ground is needed, in which poorer farmers, in particular, around the world would also be given access to the best new genetic technologies, including better seed varieties which enable crops to survive in tough and changing climates. These new varieties could enable farmers in sub-Saharan Africa, for example, many of whom were women, to reduce their fertiliser use and reliance on fresh water in an unpredictable climate. Gordon argued that the primary need in developing countries is for better training and extension services; more access to improved seed varieties at affordable prices; more support to farmers to get their products to market in good condition and at a fair price; and better access to financial support on favourable terms, with which to buy appropriate technology for their farms. By way of one example, across Africa, average maize yields are currently 1 ton per hectare: with the right mixture of fertiliser and NPK (nitrogen, phosphorous and potash), it would be possible to raise yields to 5, 6, 7 or 8 tons per hectare, from Ethiopia to Zambia, which would lead to a transformation in farmers' prospects.

Gordon was also enthused by the potential for mobile

technology to help farmers get a fair price at the market, and to know when to harvest and what the weather had in store. He had seen lots of young entrepreneurial farmers in Rwanda, Mozambique, Uganda and elsewhere find new models of leasing and updating farm technology, such as innovative corn-threshing machine businesses run by young men in Kampala. These new, IT-enabled businesses held the promise of providing a new economy for young people in rural areas, meaning that some might stay on their farms rather than move to the cities. The latest technology also helps farmers to predict extreme weather, providing insurance and assisting them to withstand the worst agricultural crises. 70,000 farmers in Tanzania are now insured against climate risk, in a model supported by the London insurance house Willis Towers Watson. The future, in Gordon's view, lies in a judicious combination of the old and new: reducing the environmental impact of farming, through better management practices, while embracing the benefits of new seeds and cutting-edge technologies.

How we manage the world's soil is intimately tied to the broader 'food system': the foods we choose to eat in a changing climate and society, and how we organise our agriculture to meet the demands of this system. To understand this larger context better, I went to speak to Dr Tara Garnett, one of Britain's foremost experts on the food system, known for her clear-minded assessment of the scientific evidence and as a debunker of myths and the ideological extremes which often arise in this area.

Soil

In a café in Finsbury Park, Tara began by saying that a good dictum for the world to live by was the American food writer Michael Pollan's famous seven words of universally applicable advice: 'Eat Food. Not Too Much. Mostly Plants.' The world's population needs to eat enough healthy food, predominantly fruits and vegetables, and not overdo it: if we organised ourselves accordingly, we would be on the right track. She then set out three further core principles to which we should aspire.

The first is that the world needs to secure and protect its remaining forests, peatlands and other critical ecosystems at all costs: further agricultural production should not come at the expense of the natural world, but instead by making existing agricultural land more productive. Secondly, we need a drastic effort to reduce food loss and waste – to which I return in Path Eight – given the staggering third of the world's food that is currently wasted. And thirdly, the world needs to work out where and how to manage its cattle, so that they can be grazed in the right areas and in the right ways to enhance the health of the soil and draw carbon back into the ground, while removing pressure off critical ecosystems elsewhere.

Tara's point about livestock prompted me to challenge her as to whether the whole world eating vegetarian (or even vegan) was the solution, such that there would be no more soy produced to feed cattle, pigs and chickens, with all the deforestation footprint this entails – with that land used instead to produce more nutritious, less water-intensive crops. Tara responded that the footprint of a vegetarian who consumes a lot of milk, cheese and eggs, albeit no meat, is not very different from someone who eats a low amount of dairy with the occasional amount of free-range meat (or fish). Vegetarians still rely on the international system currently enabling our high levels of meat consumption to take place. Instead, the challenge in her view is to eat very little meat, dairy and fish, and to ensure that the meat one does eat always comes from an ethical and environmentally undamaging source. More countries should also, in her view, follow the examples of Sweden and Norway which had established strong national guidelines urging a much healthier, vegetable-based diet.

The over-consumption of red meat is one fairly uncontroversial aspect of the global health challenge. In many parts of the developed and developing world alike, rates of red meat consumption are too high; and where this is not the case, consumers are fast seeking to replicate those same high levels. High rates of red meat in a diet can lead to higher rates of stomach and bowel cancer. The environmental consequences are also significant, in terms of water use, food production to feed cattle, deforestation as a result of cattle roaming extensively and soil compaction. Although well-managed cattle can play a role in restoring some ecosystems, the simple truth is that one of the most profound (as well as the most straightforward) steps humanity could take to reduce pressure on the soil and on biodiversity would be to cut down dramatically on eating red meat.

Bill and Melinda Gates have recently begun an effort to provide well-bred and sturdy chickens to rural African households as an alternative to red meat, and the Ethiopian government is one of a number of African countries seeking to encourage significant uptake of chicken consumption. In a world of radically less red meat production, areas in Brazil currently devoted to soya and cattle could be put to much more sustainable use, sequestering carbon in the soil and allowing biodiversity to flourish. Those same landscapes could instead be devoted to forest and ecosystem conservation and restoration, along with diverse agroforestry schemes which could provide healthy food and a diversified income to rural farmers.

Across Africa and around the world, a movement to restore soils and forests is under way. Monique Barbut, the Moroccan environmentalist who chairs the UN Convention to Combat Desertification, told me that 'the images of the earth from space show a beautiful, blue-green planet. That is the earth we inherited. What it may look like by year 2100, just 80 years from now, will be quite different, if we continue on the path we are treading of degrading, on average, 12 million hectares of productive land every year.' But our future is our own, she went on: 'We create and recreate it. Our generation is in the best position to alter the cause of humanity's destiny for the better, and at affordable cost.' The world needs to rehabilitate and restore at least 12 million hectares of degraded land every year, and of the 169 countries that acknowledge they are affected by land degradation, 110 have committed to do this by 2030. Monique believes this is possible: 'To put it simply, if each of these countries rehabilitated just 110,000 hectares every year, we could meet this target immediately and affordably.'

In early 2018, the state of Andhra Pradesh, in south-eastern India, announced it would transform its farming over the coming six years, eliminating chemical and fertiliser use, and encouraging the widespread adoption of regenerative agricultural principles and better soil management. Once an iconic state of India's original 'Green Revolution', Andhra Pradesh has now set an ambitious vision which other Indian states as well as other regions of the world can emulate: 'Zero-Budget Natural Farming'. The region intends to eliminate the use of external inputs, focusing instead on the use of in-situ resources to rejuvenate soils, as well as restoring ecosystem health through the use of diverse, multilayered cropping systems. In so doing, it seeks to improve the incomes, knowledge and welfare of 8 million farmers. Andhra Pradesh is an inspiring microcosm of the world we wish to see: a world in which we manage the soil as if we mean to stay for centuries to come.

Water

Davina and I met when we both worked for the AMAR Foundation, which was established by Baroness Emma Nicholson to assist the Marsh Arabs of Southern Iraq, who were viciously attacked by Saddam Hussein in the late 1980s and early 1990s. The Marshlands of Southern Iraq, a unique freshwater ecosystem immortalised in the pages of Wilfred Thesiger's accounts of the 1950s, and which some believe formed the landscape which originally inspired the 'Garden of Eden', were devastated by Hussein: he built dams and levees to drain the Marshes, and polluted the waterways, as part of a vicious campaign of political suppression.

By the time of Hussein's demise in 2003, many felt that it might be impossible to save the Iraqi Marshes – most of whose people had been displaced to the outskirts of Iraq's second city, Basra. In the decade since, however, local environmental leaders and communities have worked hard to recover the original Marshlands, with the support of the UN Environment Programme. Some of Hussein's dams and levees have been removed, the replanting of local species supported and fishing reintroduced in some previously decimated fresh waters; all in an attempt to restore the ecology and original way of life on the Marshes. And while the ecosystem has certainly by no means returned to the idyllic state Thesiger described – 'firelight on a half-turned face, the crying of geese, duck flighting in to feed, a boy's voice singing somewhere in the dark, canoes moving in procession down a waterway, the setting sun seen crimson through the smoke of burning reed beds' – in significant areas there have been encouraging signs of the return of fresh water, fish and biodiversity.

One of the beautiful aspects of the original Marshes was the example they provided of humanity living in harmony with, and sustainably managing, its freshwater ecosystem. To this day, there are examples of where this continues to happen. One inspiring instance in India is in the parched landscapes of northern Rajasthan, where the water activist Dr Rajendra Singh has worked with local communities to reintroduce age-old systems of using small levees at key junctures of the rivers to protect and restore water sources across the small farms and fields of these hot and arid areas. By working with the ecology of the region, rather than imposing a top-down view at odds with it, Rajendra's approach has demonstrated the potential for water sources to recover where they had previously been lost. Rajendra is sometimes now referred to as the 'waterman of India', with his approach replicated in other regions of the Indian subcontinent, and even as far afield as areas of rural Scotland.

Traditional approaches to water management are also undergoing a renaissance in parts of the Middle East and North Africa. For a long time superseded by modern irrigation, the traditional waterways of the *aflaj* in Oman – the intricate patchwork of narrow, mud-walled underground and surface canals which exist in urban areas and in agricultural landscapes – are being revived in times of water scarcity and climate variability. Communities appreciate the rules which govern the use of the *aflaj*; in rural areas, for example, households get first use of the fresh water, followed by schools and mosques, followed in turn by a portion going to communal freshwater baths, and then the rest to the fields. But the traditional systems can only survive if societies agree to abide by the rules, taking pressure off these scarce resources and polluting them less.

There are other heartening examples of where fresh water is being managed much more effectively and sustainably, combining insights from modern technology as well as ancient understanding. These, including in sub-Saharan Africa, show how communities can adapt to the changing climate's impact on scarce water resources through better community planning and the use of mobile technology. In Burkina Faso, WaterAid supports villages in improving the way they manage their water resources, by providing training in monitoring, interpreting data and sharing risk information. This information allows communities to plan which groundwater wells to use, to request seeds that need less water in the dusty and arid soil, and to build sand dams in the right places.

The temperature can often exceed 40 °C: a hostile climate, with the dry season lasting as much as eight months a year, as the *harmattan* – the hot dry wind of the Sahara – blows across the land, making rivers evaporate and groundwater drop to precarious levels. When the rains end, villagers are forced to dig for days in the river bed or to travel for miles in search of water to drink or to irrigate. In the words of one of the community leaders, Balima Kalim, WaterAid's support has transformed their lives: 'Now we can monitor the water, there has been a big change in the welfare of the community. We make graphs so we can show people the evolution of the water level. We tell them when they can start to sow their crops and what quantity of water they can fill the buckets with, so that the plants don't die. Now there is no more suffering for water, there are no more quarrels. I have hope for the future.'

In China's Turpan Prefecture, in the east of Xinjiang, agricultural and industrial growth has placed overwhelming pressure on finite sources of groundwater. Even the *karez* – the region's traditional tunnels and access shafts – have been drying up under such strain. And so, working with the World Bank, the prefecture has recently agreed a regionwide plan to reduce the area under irrigation and impose a strict water-consumption cap, which it has explained to farmers and other users across the area. Using satellite technology and remote sensing to assess evapotranspiration and monitor water use, and to reform water allocations, the region has established a means of extracting more value from less land and water, shifting its agricultural production from low-value cotton and maize to more lucrative melons and grapes. As a result, farmer income has gone up by 4 per cent, while groundwater decline has been reduced by 170 million cubic metres.

As a doubtless naive, and certainly idealistic, student at university, I learnt of the existential struggle being waged by a group of *adivasi*, or tribal people, in Central India against a World Bank-funded large dam. The 'Sardar Sarovar' was to be a series of dams spanning the states of Madhya Pradesh, Maharashtra, Rajasthan and Gujarat. The dams were designed to channel water and energy primarily to the wealthy and politically well-connected sugar farmers of Gujarat state, as well as to the industrial factories of the growing and polluted city of Ahmedabad, and were set to submerge the lands of tens of thousands of *adivasi* people living on the banks of the Narmada River, disrupting a way of life which had existed for millennia, and submerging many thousands of hectares of forest as well.

The Indian novelist Arundhati Roy was an outspoken critic of the projects. The activist Medha Patkar led the villagers' efforts to push back against the dam, and still to this day goes on hunger strike periodically to protest against how the displaced people are being looked after, all these years later.

Inspired by a university friend Keith Hyams, I travelled one summer to the villages of central Maharashtra and Madhya Pradesh to lend whatever support I could to the headquarters of the activists' campaign, 'Narmada Bachao Andolan'. In retrospect, I wonder what I had to offer these fearless and dignified people other than goodwill and a general willingness to help, but they were kind enough to welcome me, and I spent a memorable few weeks in the heart of rural India caught up in the efforts of the people to make their non-violent case. Late into the night, we sat in communal tents and activists' houses, over delicious vegetarian food, wearing an Indian *lunghi* and putting the world to rights. Many thousands of these people had been displaced by the state to precarious dwellings on the outskirts of the region's cities, and there left to their own devices, with scant compensation for the livelihoods and way of life that they had been forced to give up. Other villagers remained, vowing to be flooded (to death, if needs be) by the rising river. Others still went on debilitating hunger strikes, in the Gandhian tradition, in swelteringly hot tents in the towns, until they were taken away by the authorities and force-fed through the nose.

My summer trip to the villages, set in beautiful forested landscapes on the water's edge, culminated in several hundred of us travelling to Mumbai to march through the monsoon-drenched streets. I remember sitting in a square beneath a statue of Gandhi with the villagers, a fistful of rupees clutched in my hand, my only dry possession. We spent long, hot days in the public cricket grounds hearing speech after speech, as Medha Patkar embarked on another brave hunger strike. I remember, too, learning of the 9/11 attacks from a newspaper on the street corner, and feeling that the world had shifted on its axis.

Medha Patkar came to Oxford the following summer and I became her secretary and scribe, firing off letters from early in the morning to late in the night at the foot of her bed in Jericho. I still feel grave misgivings about many of the world's large dams, especially when they lead, as Sardar Sarovar did, to such huge waves of displacement of people, alongside submersion of vital forests and, in Turkey's case, archaeological ruins. Alas, the efforts of the 'Narmada Bachoa Andolan' have not prevailed: the dams were finished and officially inaugurated with much fanfare on Prime Minister Narendra Modi's sixty-seventh birthday in 2017. The project was an 'engineering miracle' and a 'symbol of India's new and emerging power', he said. The long-term resettlement of the thousands of affected families remained unresolved. The politics of fresh water – how it is managed, who benefits, the inevitable trade-offs between different uses of water – have stayed with me, and form a backdrop to our fourth path to a hopeful future.

With a dramatic reduction in greenhouse gas emissions through renewable energy, our forests protected and our soil replenished and restored, the world's precious and finite freshwater supplies are more likely to withstand the pressures of the centuries to come. For the stark truth is that climate change is already having a significant impact on the world's freshwater flows, making them more variable and unreliable, with significantly more droughts and floods across the world. At the same time, humanity is placing ever greater strain on water supplies to meet growing demand for energy, agriculture, cities, infrastructure and dams (which continue to be frequently built in South-east Asia, sub-Saharan Africa and elsewhere).

Forest loss and soil erosion negatively affect the quality of water, as do sewage and waste from cities. More and more rivers are gravely polluted and do not run to the sea; and the risk of conflict in and between water-scarce countries and regions, historically surprisingly rare, is set to increase. The world urgently needs to secure, restore and protect its vital sources of fresh water, particularly – as Professor of Biodiversity and Ecosystems Georgina Mace of University College London, explained to me – in some of the most important strategic water basins, such as the Himalayas, the Mekong and the Nile, on which many hundreds of millions of lives depend.

'Water is the driver of Nature,' wrote Leonardo da Vinci; 'by means of water, we give life to everything', as the Koran puts it. The good news is that the world is becoming more and more aware that water is a fundamental prerequisite of social and economic development as well as of environmental sustainability. We are more conscious than ever before of the need to live with a lighter water footprint, to pollute water a lot less, to capture water run-off, to use better irrigation systems and to protect important freshwater catchments. The human dimensions of the water challenge are also striking. Some 844 million people in the world still do not have access to safe drinking water, and 2.3 billion people lack access to basic sanitation. Throughout Oli's first year, I have been constantly reminded of how incredibly fortunate we are to turn the tap on to receive clean drinking water, to be able to run a bath, to flush a toilet, to use a dishwasher and a washing machine, and to water our garden.

As Barbara Frost, then CEO of WaterAid, the British water and sanitation charity, told me: despite all the progress of recent decades, the lives of many hundreds of millions of people are still thwarted daily by their lack of access to fresh water, clean toilets and sanitation. Many millions of people still have to walk for many miles to get water; girls have to leave school due to the absence of proper toilets; and floods in Bangladesh and elsewhere periodically lead to the contamination of freshwater supplies with human sewage. Across many of the world's countries, governments still have a long way to go to deliver the investment, governance and infrastructure needed to provide fresh water and sanitation to all their people, in particular vulnerable urban and rural communities.

When I was 18, fresh out of school, I went to live in a Tibetan monastery in the foothills of the Indian Himalayas: Kirti Monastery, in Dharamshala, in the fertile state of Himachal Pradesh, where apples and tea are grown in beautiful orchards. In 1959, the Dalai Lama left Tibet and embarked on his dramatic journey across the Himalayas – a journey which Tibetans fleeing Chinese repression continue until this day – and had settled here. Dharamshala is now home to the Tibetan Parliament in Exile and receives many thousands of well-wishers every year, of which I was (and remain) one.

I taught English in the monastery to a variety of classes, ranging from novice monks, aged nine, to scholarly adult monks who already spoke and read English well. I woke to the sound of the monks saying their prayers at dawn, and we ate together: piping hot Tibetan tea, made of butter and milk washed down with warm steamed bread. We watched the sun rise over the Himalayan hills, with the landscapes and morning dung fires of Himachal Pradesh unfolding on the plains below. On special occasions, the monks and I ate *momos*, meat dumplings; Tibetan Buddhists (even if born in exile) are rarely vegetarian due to the importance of yak meat, milk and butter to ensuring energy and survival in such a cold and forbidding landscape.

On Saturdays, one of my closest monk friends, Kathup, would lead us on an expedition up into the mountains to a freshwater spring (next to which a small Hindu temple had been built) where we swam and washed our clothes with a small bar of soap, the rocks adjoining the stream covered in the monks' red robes laid out to dry. On Sundays, the other English teachers and I were among the few to attend a church service in the damp and neglected nineteenth-century Anglican church, led by a friendly Keralan priest who served his modest congregation sweet white bread and even sweeter white jam, perhaps as a reward for coming.

One of my most brilliant, gentle students, Dawa, from Amdo, the north-east region of Tibet, was a gifted poet in the Tibetan tradition, his verse greatly appreciated by the Dalai Lama and by his peers. Dawa was kindness personified: talkative, with a broad and winning smile, and always welcoming and hospitable. Some time after I left, and to my great sadness, Dawa died in captivity in a Chinese-run prison in Tibet. His mother, back in the Amdo, had been unwell, and wrote to Dawa in exile. Dawa decided to risk returning to Tibet to see her and took with him, hidden in his luggage, a photograph of himself receiving a poetry prize from the Dalai Lama he wanted to show to his mother and family. Tragically, Dawa was apprehended on his return; his bag searched; the photograph found. He was put in prison for this supposedly seditious possession, where he died – we can only assume beaten and tortured. I learned about this in 2002, and with the help of Free Tibet and Amnesty wrote up Dawa's case, but justice has never been done.

It was in that same year, 2002, that I first met the writer and journalist Isabel Hilton, who is deeply knowledgeable about Tibet and China, and who had kindly come to give a talk to the Oxford Free Tibet Group that I ran for a while. She is a seasoned observer of China since the 1970s, a fluent Chinese speaker and the editor of the bilingual 'chinadialogue' for the past decade. I went to see her in chinadialogue's office in trendy Hoxton to ask her for her thoughts on how the Chinese were dealing with the multiple environmental challenges they face, including fresh water in particular.

Isabel was characteristically eloquent and sanguine, and began by speaking about the Yellow River, central to Chinese civilisation and to its conception of itself as a nation, which had faced a crippling drought in 2007. Whole parts of the estuary had dried up. This had come as a major shock to the Chinese and served as a startling wake-up call. Since then, pressures on the Yellow River have only intensified, and a changing climate is continuing to have a significant impact on the river – although some conservation measures have been adopted, with some success, such as downstream beneficiaries of improved freshwater paying farmers upstream to stop deforesting.

Increasing incidence of drought coupled with excessive aquifer use was also compounding water scarcity in urban areas of China. And while the Chinese were excellent engineers, this has frequently translated into excessive engineering zeal: no country had built more dams, especially during the Mao era, which sought to dominate nature rather than peacefully coexist with it. The current South-to-North Water Diversion Project – the largest transfer of water between river basins in history, set to move as much as 45 billion cubic metres per year – is another case in point. Encouragingly, though, President Xi has been the architect of a national commitment for China to become an 'Ecological Civilization', and this thinking has permeated the country's Five Year Plans and decision-making. In the absence of US environmental leadership in the Trump era, there was some hope that President Xi would take the commitment more seriously than before.

Isabel noted there had been a recent improvement in Chinese concern for animal welfare, helped by increased pet ownership and public awareness campaigns against eating dog, bear bile extraction and the like. Nevertheless, there is still a long way to go, and rates of consumption of shark-fin soup and ivory are still catastrophically high, while factory farming is on the increase. The Han Chinese have a very different conception of landscape and nature than the Tibetans, and are busy building railways and dumping nuclear and radioactive waste across the Tibetan plateau (although one of the world's largest solar-power installations is also to be found in Tibet) – to the detriment of the freshwater sources to be found there, in the heart of the Himalayas.

Meanwhile, the Chinese are also leaving a significant footprint elsewhere in South-east Asia, Africa and South America, whether by building roads or oil installations in the Amazon rainforest or through supporting dams, logging and mining in some of Africa's most iconic landscapes of great biodiversity. Nearer to home, they are massively invested in an infrastructure project to cross Central and South-east Asia, the so-called 'Belt and Road Initiative' ('One Belt, One Road'). In 2017, President Xi announced that it would be 'green, low-carbon, circular and sustainable', although there are early indications that much of the investment in Pakistan, Sri Lanka and Bangladesh appears to be in coal power stations. An active dialogue is under way with the Belt and Road Initiative to support the fulfilment of President Xi's vision and to mitigate some of its environmental damage, including on precious supplies of fresh water both at home and in the neighbouring countries it will affect. A lot hangs on how this initiative develops in the years ahead and there is some hope that pressure from partner countries coupled with increasing levels of domestic concern might shift the initiative in a more sustainable direction.

Ultimately, governments, companies, communities and individuals must take seriously their role as custodians of available fresh water, allocating and enforcing fair shares among different users (and allowing for seasonal variation, climate change and the needs of ecosystems). There also needs to be a heightened sense of responsibility for proper, long-term water custodianship at the level of entire riverbasins: no amount of local action, or better management at the scale of an individual farm or factory, can replace the need for a holistic, bird's eye view to be taken across the whole freshwater system. This is by no means easy, and will require sensitive diplomacy, transparency, good will, principles of equity and the best of human intelligence. It is made harder still when riverbasins span countries, as they do in the case of many rivers including the Nile, the Tigris-Euphrates, the Mekong and the Brahmaputra. In the case of each of these rivers, it will take transboundary water management of the highest order to ensure cooperation rather than tension between Egypt, Ethiopia, Sudan and South Sudan, in the case of the Nile; Turkey, Syria, Iraq, Iran and Kuwait, when it comes to the Tigris-Euphrates; China, Burma (I continue to use the original name, as I consider 'Myanmar' to be a name chosen by the military dictatorship), Thailand, Laos, Cambodia and Vietnam, for the Mekong; and, perhaps most strikingly of all, given the numbers of people potentially affected, between India and China in the case of the Brahmaputra. Our fourth way to save the world is to manage freshwater as if our lives depend on it, which they categorically do.

Biodiversity

Two particular landscapes in Kenya fill me with hope, the Maasai Mara in the south and the Rangelands of the north. Although neither is without its challenges, both show how it is possible for humanity and biodiversity to live side by side, and for local communities to benefit from conservation.

Davina and I went to the Maasai Mara in 2013. There, in a series of memorable drives, we saw dozens of elephants, a solitary, rather nervous and forlorn rhino (the poaching crisis having taken a real toll), and a number of cheetahs, lions and giraffes. The giraffes gambolled elegantly and prehistorically into the distance in the pink morning sun; the elephants blended effortlessly into the landscape near a stream, their young drinking water, surrounded by the adults. The lions were dozing on a warm rock, languorous and magnificent, their cubs playing by their side. The two cheetahs we saw were on a hunt, weaving their way through a crowd of onlookers. We later saw three young lionesses with bloodied mouths happily devouring a recently fallen wildebeest:

nature in the raw. In the evenings, we stayed at Maasai camps in the heart of the landscape, eating mutton stew and sitting by an open fire beneath a starry sky. On the last day of our trip, the Maasai leader took us to his village, where we had tea with his wife and bought some of her woven fabrics. Just as we were leaving, he realised a goat was giving birth, in his pen: quick as a flash, he rushed in, assisted with the birth giving the little lamb a vigorous kiss of life to get it breathing, all without a moment's thought. We left on very good terms, having witnessed this auspicious occurrence.

In northern and coastal Kenya, over 4.5 million hectares have been brought together under better management to protect biodiversity and provide sustainable livelihoods for local people. Some of these dry areas in the north are home to 18 of the 20 poorest constituencies in the country, with up to 97 per cent of the population living below the poverty line. The area also suffers from high levels of violence and conflict, driven by local actors as well as regional illegal armed groups and militia. For several decades, a charity called the Northern Rangelands Trust has worked with local communities to improve their livelihoods and act as a bulwark against these forces. They have focused in particular on coming to a shared agreement on farming and livestock management, reducing illegal hunting and respecting the animals' historic migration patterns, ensuring thereby the coexistence of the natural and human worlds. Whereas to hunt lions was once held to be an important rite of passage for a young Maasai warrior, now the communities are protecting lions and scare them away from their encampments by non-violent means. Ecotourism and conservation now benefit over 10,000 local people and ensure some 3,000 students receive school bursaries. Elephant and other species numbers are on the rise. While there are inevitable tensions, the work of the Trust is worthy of replication.

Another encouraging development in biodiversity conservation is occurring in China, Vietnam, Laos, Cambodia and across South-east Asia, where charities such as WildAid and World Wildlife Fund (WWF) are using public information campaigns to make significant dents in demand for elephant ivory, rhino horn, pangolin and other species currently traded illegally. Similar campaigns, involving popular film and basketball stars, have led to dramatic reductions in shark-fin soup consumption, with young people telling their parents they do not want it at their weddings.

Until these campaigns began, many people did not know the true impact of consuming these products, and were amazed and appalled to find out that elephant tusk and rhino horn are not renewable resources, but instead involve the indiscriminate slaughter of these astonishing animals. Hard-hitting publicity campaigns coupled with savvy social-media use convey that it is, for example, not 'cool' to drink rhino horn as a hangover cure, as has been the case among the wealthy in Hanoi. The traditional medicine community in China had also disavowed the use of these products in authentic Chinese medicine, at least until a startling setback at the end of 2018 that saw this ruling overturned. To step up the pressure in a desperate attempt to reduce the poaching crisis with immediate effect, countries such as China and the UK have both adopted stringent domestic ivory bans, seeking to send a strong signal to the global market. While it is far too early to claim a victory for elephants and rhinos, given that the poaching crisis continues unabated in many spectacular parts of Africa, there are nevertheless encouraging signs that the level of action to address demand, coupled with increasing amounts of on-the-ground protection, may mean that there is still a hopeful future for these iconic species.

Biodiversity is vital for the agricultural system and for the future of global food security: indeed, there can be no true and lasting food security without due care for biological diversity. Global seed banks, such as Kew's Millennium Seed Bank and the Svalbard Global Seed Vault, are vital linchpins for the preservation of the genetic diversity of seeds in perpetuity. Svalbard is a seed-storage facility situated in the permafrost of Norway, built to stand the test of time, including natural or man-made disasters. The vault stores backups or 'duplicates' from the world's seed collections. The theory (and the hope) is that the permafrost and thick rock surrounding Svalbard will remain frozen for ever, even if power supplies were to be cut to the vault; although a flood in the permafrost in 2017 sent a worrying signal that nowhere is entirely immune to the changing climate. The vault currently holds 890,000 samples, from almost every country in the world. These range from unique varieties of Asian and African staples such as cowpea, rice, maize and wheat, to South American and European varieties of barley, potato, lettuce and aubergine. The seeds are stored at the optimal temperature of -18 °C, in custom-made three-ply foil packages in a low moisture environment. It is a remarkable operation and commitment to the future.

During a recent visit to Ethiopia, I learned of Bioversity International's work with beer and pasta producers to use endemic varieties of durum wheat. Given that Ethiopia is home to some of the world's original strands of durum wheat, it is surely only right that the country draws on them, rather than the homogenised, bland GM wheat that is used in the bulk of the world's mass lager production. Supported by the Royal Botanical Gardens of Kew, research institutions are partnering to protect the world's diversity of seeds in seed banks in India, Ethiopia, Indonesia, Nepal, Thailand, Mexico and elsewhere.

Ethiopia is of particular interest, as it is home to many of the original varieties of crops, including coffees grown uniquely in forest areas, such as the idyllic region of the Bale Mountains. A celebrated Ethiopian scholar, Professor Sebsebe Demissew, has for several decades been documenting the country's astonishing and varied flora and fauna and is an eloquent voice calling for the protection of its biodiversity. When we spoke, Professor Sebsebe expressed the hope that the Ethiopian government would plan its development with especially important areas of biodiversity in mind. He is not anti-development, only insistent that it must be carried out in a way which is respectful of Ethiopia's biodiverse landscapes. Ensuring better land-use planning and infrastructure development is a critical prerequisite for biodiversity conservation in the twenty-first century and beyond. There are encouraging signs that biodiversity is being taken into consideration more than ever before, but we have a long way still to go.

My love of biodiversity began as a child: aged five, I was completely obsessed with sharks and whales, with large collections of books and stickers on the subject, and – I am told – a fairly detailed knowledge (now, alas, long forgotten) of many different species. One of my first memories is of being upset about the existence of a stranded whale on a beach in Amagansett, Long Island, on the East Coast of the US, the stench of the rotting flesh carrying far across the sun-baked sand dunes. Slightly older, I fundraised for WWF in my free time at school, enthused by my environmentalist Uncle Adrian – although I was apparently perplexed and somewhat disheartened to learn that all of the funds had to be given to the charity, rather than a portion pocketed for personal consumption.

Tragically, the world is currently losing much of its remarkable biodiversity as a direct result of human action and decisions. Many scientists agree that the rate of loss means that the world is either already undergoing or fast approaching its 'sixth extinction' – the sixth time in the history of the planet in which many millions of species are irrevocably lost. While previous extinctions were not our fault, this one is. We are direct instigators of, and witnesses to, the loss.

Why is it that the world's biodiversity, a term first coined by the scientist Thomas Lovejoy, is so important? Michael McCarthy – whose book *The Moth Snowstorm: Nature and Joy* is an eloquent paean to the natural world – explained to me that it is the intrinsic value of the breadth, diversity and beauty of nature, as well as its capacity to engender awe, that should inspire us. Michael makes his arguments with emotion and tenderness. The 'moth snowstorm' his book refers to was the swathe of moths which would land on his family car's windscreen when he was growing up in post-war Britain, driving through the English countryside on a summer's day: now, one barely collides with a single moth on a similar journey, as they've largely been lost, mainly as a result of excessive pesticide use in agriculture.

The Prince of Wales once said in a speech that 'humanity is less than humanity without the whole of creation'. We were at St James's Palace for a meeting on the illegal wildlife trade, and I felt the room become pensive and still at his words. We are all gravely diminished by the loss of biodiversity; it is simply not right, at some fundamental level. The loss of the last male northern white rhino, named Sudan, in March 2018, struck a chord globally. The actions of mankind could lead to the loss of half of Africa's bird and mammal species in the twenty-first century, according to a recent study, if we do not take concerted action.

All is not lost, however: there is still so much biodiversity to cherish and protect, and we know with abundant clarity what needs to be done. There are many heartening examples of biodiversity conservation around the world, and these approaches need to be dramatically scaled up to save the world's most iconic, and, even more importantly, hitherto unknown, species. Fortunately, rates of global awareness have never been higher.

When I met David Attenborough, he described how 'television is very good at lighting enthusiasm and sparking excitement, although books are much better at teaching.' Television provides a 'very vivid picture of reality that has brought a recognition and familiarity about the natural world back to more than half the world's population cut off to a greater or lesser extent from it'. He also observed that the number of people involved in conservation work is greater than ever before: worldwide, we are more aware of 'the variety and beauty and splendour of the natural world' and more likely to recognise its value. 'People won't protect things unless they know what they are, and if they don't love them,' he said. Levels of awareness have increased, but so too ironically has the size of our modern economies that play such a destructive role.

I asked David to describe some of the memorable experiences he had had filming the world's natural species. He recalled one time:

. . . in northern Australia, when we got up before dawn to hide on the end of a lagoon and a billabong while it was still dark, and so the animals and birds didn't know we were there. As the sun came up in the billabong, amid the egrets and parrots and crocodiles and kangaroos, it was not yet bright enough for us to film and there wouldn't be enough light for another half an hour, so there was nothing to do other than immerse yourself in that context. The sad thing is you're not part of it, but a privileged eavesdropper; although there are of course moments when you do come to be a part of it. Sitting with the gorillas, when they take you on their own terms, they acknowledge you, and you are not hiding, and you are allowed to react in a way which you hope is amiable: these are two kinds of moments which one looks back on.

Our conversation about biodiversity continued. I wondered what, if anything, still surprised him. He responded:

The variety of nature. It is almost infinite – although it can't be infinite in a mathematical sense – but beyond the compass of any one human being. There are things we discover all the time. Just recently, a couple of years ago, there was discovered a little puffer fish in Japan, the males of which create ridges in the mud on a circular basis like a huge dahlia, metres across the muddy floors of the bay. The heart of the flower is a way of inducing the female to deposit the eggs. He spends days and days on this: how he knows what the overall pattern is, I have no idea.

101

Path Five

The intricacy and beauty of evolution is indeed consistently moving to behold.

In 2000, I visited the Galapagos Islands. I still remember the joy I felt during a close encounter with the blue-footed booby, and understanding how this bird was living proof of Darwinian evolution. The albatrosses were in full mating season, in an elaborate and wholly enchanting courtship involving protracted bouts of jousting with their beaks, a kind of amiable and intricate form of sword fighting, and seemingly oblivious to the small group of human onlookers. On another island, I sat on the beach away from a gathering of lazy seals; and snorkelling in the sea, amid coral reefs bright with colour, I swam with sharks, dolphins and penguins, the penguins fizzing past like turbo-charged mini-submarines.

Even then, local conservationists were worried that too many tourists were coming to the islands, and that too many people were also living in the main town, leading to a significant impact on the archipelago's fragile ecology, through waste, pollution and the introduction of goats and mice. The situation has deteriorated since then, with increased numbers of tourists far beyond the islands' capacity to cope. I have always been haunted by the idea that we humans can end up inadvertently ruining the places we love. I deeply hope that this never happens to the Galapagos; efforts such as those of the Galapagos Conservation Trust to encourage better management and care deserve every support.

Another set of challenges besets the extraordinary national park in the Democratic Republic of Congo, Virunga. I was fortunate to meet its inspirational director, Emmanuel de Merode, who had survived being shot four times in the legs and stomach by unknown assailants when driving back one night from Goma to the park's headquarters. Emmanuel runs a team of 550 rangers, responsible for the surveillance and protection of some 3,000 square miles of the national park. In the past ten years, 160 of his rangers have died, a number which is tragically representative for the ranger profession globally, as the charity the Thin Green Line Foundation eloquently points out. Virunga spans the Albertine Rift, in the heart of equatorial Africa, a region characterised by mountains, great lakes and much biodiversity. It is also home to the Rwenzori Mountains, central savannahs full of hippos, elephants and lions, and – in the south – mountain gorillas which inhabit the remaining rainforests on the slopes of several dormant volcanoes. On the borders of the park, several million rural people make their living from farming. Former and current combatants from the conflicts between Rwanda, Burundi and Uganda also live in or traverse the park. There are significant oil and mining interests. Emmanuel spoke passionately about the beauty of the park, and the commitment of his rangers, as well as his hopes that new financial models for ecotourism and fishing in Lake Edward could provide a meaningful alternative livelihood for local people. In May 2018, the park was closed for a year, after 12 rangers were killed and two British tourists abducted. Emmanuel has instructed a thorough security review to be undertaken. One can only hope that this remarkable park can withstand the conflict and the human pressures which beset it, seemingly from all sides, and that people can once again come to learn and marvel at the biodiversity it continues to contain.

Alexander von Humboldt was one of the great explorers of the nineteenth century and an early witness to the biodiversity of the Andes and the Amazon. Humboldt explored the landscapes of South America with astonishing curiosity, describing hundreds of species with great care and eye for detail. Around the world, today, scientists follow in his footsteps to document and understand the natural world. One of Humboldt's modern-day successors is Professor Kathy Willis, outgoing Director of Science at Kew, who leads a team of 380 scientists in Kew's herbarium and fungarium, as well as its Millennium Seed Bank. Between them, they curate and research some 8.5 million specimens. Kathy and I met one spring day in Kew to discuss her work. She began by praising this 'extraordinary national asset in West London', describing the history of its collections since 1759 and over 150 ongoing projects with 400 partners.

Their purpose is a simple one: 'to understand the diversity and distribution of plants on earth'. Over the past 20 years, Kathy's thinking on biodiversity has evolved. Originally, she explained, she considered biodiversity first and foremost as 'a precious resource, the evolutionary engine house of the world'. The sheer diversity and distribution of plants is 'as important as any other aspect of life on earth'. Her arguments for biodiversity protection were then based around the intrinsic value Michael McCarthy referenced earlier: 'nature for nature's sake'.

More recently, Kathy has felt that these arguments haven't worked: rates of extinction and land-use change remain 'terrifying'. And so she has been forced to consider again 'how the world really works'. She was particularly struck by a journey to Madagascar, where in villages she visited: 'poverty overrides concern for a beautiful tree, if you need the charcoal to feed your children.' And so, Kathy felt, the challenge was to quantify and understand the value of the tree for 'the services to humans it provides', and thereby to show how biodiversity is essential to human life and well-being. If this value could be demonstrated, and an economic system established in which the community would stand to benefit from keeping those trees standing, then biodiversity would be better protected.

Around the world, Kew is seeking to help people find solutions to pressing environmental problems. For example, where biodiversity is being lost due to agricultural expansion, Kew seeks to find and draw attention to alternative crops with high yields and high protein content which could be used as a substitute for the world's main crops (themselves at significant risk due to their limited gene pool). *Ensete ventricosum* – from the banana family – is one such example. It provides food to

20 million Ethiopians. The plant reproduces vegetatively and flowers in the wild. 'The world needs this genetic diversity and resilience,' Kathy argues, before describing Kew's work to protect and document different subspecies of the *Ensete* plant.

In another example, Kew has helped Ethiopia to draw on its herbarium records to establish the right climate for coffee, assess the areas it currently grows in and determine the risks posed to coffee by climate change. Kew's research shows that climate change, left unchecked, is likely to have a negative impact on between 80 per cent and 100 per cent of the existing area of coffee production in Ethiopia. Instead of 'doom and gloom', however, Kew created the *Coffee Atlas of Ethiopia* which identified 130 species of coffee with greater climatic tolerance, and that are 'quite happy at 40 °C'. One variety in Madagascar is a coffee bean the size of a peanut, which could be produced in Ethiopia if the climate changes as the models predict. The challenge globally is to find alternative plants and climates where new species can flourish.

'Plants are much smarter than animals,' Kathy reflected. 'If you went around a zoo knocking the animals on the head with a hammer, they would all die. If you scythed down, or burnt, a botanical garden, ten years later – given the right environmental conditions – its seeds would grow underground and germinate.' The Millennium Seed Bank is a 'living seed bank', where the curators check seeds to see if they are still viable, germinating and growing them on. It draws its inspiration from the original seed bank in St Petersburg, established by the celebrated Russian botanist Nikolai Vavilov, which was staunchly defended by its heroic staff during the siege of Leningrad in the Second World War, a number of whom starved to death underground while protecting the seeds, rather than eat the seeds under their care.

In summary, Kathy calls herself 'a proud optimist': while of course there are terrible losses, there is also lots of evidence that the pendulum is swinging. People are much more aware than ever before. 'In Brazil, the reduction in deforestation rates is significant, while Colombia is making great progress; much of Europe is undergoing reforestation.' She was also encouraged by how big businesses and major players on the global platform are also stepping up to the plate: 'they have rights to huge swathes of the earth, and previously only spent money on symbolic environmental impact assessments, whereas now they realise biodiversity is better for their profit lines, due to its role in providing clean water, healthy landscapes, maintaining soil and reducing soil erosion.' Biodiversity is an 'incredible natural asset', she concluded, and the world is beginning to take notice.

As we have seen, wherever you look, and despite all the odds, communities, scientists and governments are seeking to protect biodiversity and the earth's wonderful variety of species and ecosystems. There are also some significant efforts under way to map these species better and to document the world's key areas of diversity more systematically and at a greater level of detail than ever before. But the community concerned with these matters has perhaps, for some time now, lacked a striking political narrative or 'story'. Efforts to promote and draw attention to the economic benefits of nature have gained some recent traction with policymakers, but have not yet transformed the way the world looks at biodiversity. And while we have achieved strong progress on the climate, biodiversity has fared less well as a pressing international concern and campaigning issue.

The year 2020 is critical for biodiversity, as world leaders will meet in China to assess global progress towards meeting its biodiversity goals. China's presidency is a strong opportunity to get countries to sign up to, and deliver on, a more ambitious 'global deal for nature'. It is also a chance for societies to adopt a long-term goal for nature, just as we did for the climate. We need a new rallying call for biodiversity, equivalent in scope and magnitude to the push for climate change action; and we need it fast. Biodiversity must not be the forgotten environmental issue of our time; instead, it must be at the forefront of the global effort from here on.

Ocean

From the late 1500s onwards, the Grand Banks of New-foundland were one of the most prolific and celebrated of all fisheries, providing livelihoods for thousands of fishermen from near and far. By the end of the twentieth century, however, stocks of cod had all but vanished, the result of chronic overfishing. To address the crisis, in 1992, the Canadian federal government placed a mora-torium on cod fishing, which led to devastating social consequences, including the lay-off of many thousands of workers, fishermen, boat builders and others associ-ated with the fishing trade. The impacts on local com-munities were long-standing, with many people leaving for the cities, and abiding poverty and dislocation for those that remained. However, over the ensuing period, while the moratorium on commercial fishing remained in place, recreational and artisanal fishing continued. And as cod stocks began to stage their miraculous recov-ery, the prospects for shrimp and crab also flourished – and the value of the fish caught in the area over time went up. The government established a Sustainable

Fisheries Framework to ensure that stocks could be restored, based on a combination of marine protected areas, sustainable harvest strategies, strict by-catch policy and the introduction of aquaculture. The recovery of the Grand Banks, while not yet complete, is an iconic story of great promise to the world: even a fishery as decimated as this one, managed more sustainably, can return to good health. In the marine environment, nature, given half a chance, can recover and flourish.

The Pacific nation of Palau is made up of 250 islands, home to 1,300 species of fish and 700 species of coral, as well as to one of the world's largest tuna fisheries (65 per cent of the world's tuna comes from the Pacific Ocean). Unregulated, unreported and illegal tuna fishing deprives countries in the Pacific region of about $1 billion every year in terms of fishing revenue. By-catch is another big problem here: one-third of the tuna haul is accidentally caught fish, such as silky and blue sharks, rays and turtles, which are caught up in the fishing vessels' long lines of bait. This is not only an ecological tragedy but also a terrible waste. In response, Palau's fishing fleet has worked with environmental organisations to adopt more suitable fishing techniques, gear types and bait. A simple change in hook shape and the use of fish rather than squid baits has led to a significant reduction in by-catch, with the fish bait more likely to dissolve in a turtle's mouth rather than it getting caught on the hook. Palau has also established a shark sanctuary and banned the highly destructive practice of 'bottom trawling' in its waters.

Overall, 80 per cent of Palau's waters – some 193,000 square miles – are destined to become a marine reserve by 2020, with its fisheries either well managed or left to recover, in accordance with the traditional Palauan notion of *bul*, which lets depleted or overfished reefs replenish naturally. A ban on underwater mining has also been put in place. Palau's 21,000 people stand to benefit from the long-term commitment they have made, and will hopefully be rewarded by a market that is increasingly committed to sustainable sourcing. And they have the world behind them, in the form of marine NGOs and in some instances technology firms too: a company, Satellite Applications Catapult, is supporting the enforcement of the new area, drawing the authorities' attention to illegal fishing boats.

The Pitcairn Islands, which belong to the UK, are another hopeful example of a spectacular ocean ecosystem. In 2015, the UK established the world's largest continuous marine reserve in the waters surrounding the islands, banning commercial fishing from these waters; although even here – on the verge of Antarctica – plastic particles have been found at the bottom of the ocean. These islands are a remarkable treasure trove, home to the South Pacific's deepest corals, amazing endemic species and biological treasures, including whales, sharks, fish and laughing gulls. The ecosystem is under pressure, however, from an industrial krill fishery nearby that could disturb the food chain in the entirety of the islands.

In early 2018, TV viewers were shocked to see a parent albatross giving pieces of discarded plastic to its young in David Attenborough's series *Blue Planet II*. An edition of *National Geographic* carried the arresting cover of an iceberg above sea level, morphing into a plastic bag beneath. Seemingly for the first time, millions, if not billions, of people appeared to gain a visceral understanding of the impact of plastic. While so much of the ocean is unobserved, and so much of plastic's impact takes place at the microscopic level, here was stark evidence in plain sight. And in today's connected world, millions of people moved by an issue can make their voices heard.

In the months that followed the release of *Blue Planet II*, and other programmes like it on Sky, the British Prime Minister Theresa May and her Environment Secretary Michael Gove both made a series of commitments to address plastic waste and to increase the UK's level of funding and conservation in its overseas marine territories. Improbably, the Prime Minister even raised the issue of ocean plastic with President Xi Jinping during a state visit to China, going so far as to give him a copy of *Blue Planet II* on DVD, and encouraging China to do more to address its role in the global plastic crisis.

The reappearance of the ocean on the global agenda in a serious way comes at a vital time. For there are other pressing threats to the ocean, all of which require a concerted and systemic international response. Over-fishing is the first, causing great damage to the marine environment, as well as to the long-term livelihoods of fishing communities. The potential collapse of so many of the world's fisheries – with as much as 85 per cent of the world's commercially harvested fish stocks currently judged to be at breaking point – represents one of the most significant food security, human development and political challenges of our time. Nearly half the world's population – some 3 billion people – depend on fish as their primary source of protein. When fish stocks are managed poorly, the fish may not be able to reproduce and recover sufficiently to survive.

We can turn this situation around: with political leadership and community buy-in, it is possible to implement a more sustainable fishing regime based on the recovery of the stock. While such a management plan can lead to a short-term loss for the fishing community, and these losses should be foreseen and compensated for, a better and more sustainable fishery, if properly managed, will lead to long-term gain for fishermen and their communities. Across the world, there are many inspiring examples of recovering fisheries, as we have seen in the Grand Banks.

The next challenge is to address the ocean's warming temperatures, which reduce its ability to absorb carbon and which may also be acting as a catalyst for ocean acidification. These changes in the chemistry of the ocean have significant impacts on other organisms, such as oysters, mussels and starfish: the more acidic ocean waters become, the more they inhibit these species' growth, with knock-on negative impacts for the ecosystem and the food chain. Even if we were to stop all global greenhouse gas emissions today, it would take around 75 years before the effects of ocean acidification were fully to reverse. And so we must make every effort we can to bolster the resilience of the ocean to these warming temperatures by improving its health and reducing the strain we place upon it.

Then there is the question of the world's precious tropical coral reefs, of which 50 per cent have already been lost over the past three decades. There are real concerns in the scientific community that corals simply cannot survive the likely changes in our climate foreseen for this century; even a 1.5 °C change in global temperature will lead to the loss of 70–90 per cent of the world's coral reefs. Coral bleaching has already led to the widespread death and decay of this special ecosystem. From a typical interval of 30 years in the last century between bleaching events, the gap now is only six years. Given that 'cool' La Niña currents are now warmer than the 'warm' El Niño currents of the past, the vital period of regeneration that the cool cycle once brought is now no longer long enough to enable coral reefs to recover. Meanwhile, plastic pollution also brings pathogens to the reefs, with the risk of disease to corals exposed to plastic pollution increasing by as much as twenty-fold. This is a disaster for the reefs, which are already suffering from the impacts of warmer and more acidic waters as well as the increased pressure of fishing.

The only systematic solution to the threat posed to the reefs is comprehensively to mitigate climate change. A massive year-on-year reduction in greenhouse gas emissions – starting now – is required to keep global warming to within 1.5 °C or 2 °C. Ultimately, to achieve this is the only proven and robust way of ensuring that at least some of the reefs will continue to flourish. But there are other, more local measures which should be taken – including community-led as well as nationwide attempts to reduce marine pollution and to carry out beach clear-ups, as well as scientific efforts to replicate healthy corals in laboratories and then return them to the wild that can help to mitigate the problem.

Finally, pollution of the seas from fertiliser run-off, industrial effluent and human sewage from cities, and petrol and waste from shipping, are all also exerting a huge toll on the marine environment. Most of these problems, with the exception of shipping, can only fundamentally be addressed through better waste management on the land. They will all require a major effort to be overturned in the years to come.

Despite this daunting set of challenges, we must not lose hope: there is still time to turn things around, and the capacity of the ocean to restore itself to good health remains high. There are also still remarkably large areas which have so far managed to withstand the pressures humanity has placed upon it. And so it is that a healthy, recovering, resilient ocean is still firmly within our grasp, and would benefit the billions of people who depend on the ocean, as well as its many ecosystems and species.

The world needs to expand its moral compass, political focus and collective willpower to bring about the same level of lasting change in the marine environment that we aspire to see on the land.

Fortunately, a combination of well-enforced marine protected areas, better governance of the high seas, a transition to more sustainable fisheries, and addressing ocean acidification and marine pollution, makes very strong economic as well as environmental sense. From the Seychelles to the US, the UK to the EU, and from the so-called 'Small Island Developing States' (which increasingly refer to themselves as 'Large Ocean States'), whose very existence is threatened by climate change, to the waters of South America and South-east Asia, there are dozens of encouraging examples of better marine management taking place and a greater awareness of the ocean's predicament than ever before. A 'new deal for the ocean' is possible, but it will require sustained political leadership, as well as a sense of shared responsibility and action from governments, companies, scientists, individuals and campaigning organisations. There really is no time to lose.

To find out more about the world's efforts to protect the ocean, I began by seeking out Simon Reddy, a former Political Director of Greenpeace who subsequently became Executive Director of the Global Ocean Commission. The Commission, which was housed at Oxford University, was a three-year project to put the ocean on the international political map, in the run-up to the negotiations around the Global Goals and the Paris Agreement.

We met on a winter's day in late 2016, and began by discussing how the ocean had always been lost in a 'spaghetti soup' of some 576 overlapping international treaties and instruments. While all of these have some bearing on the ocean, none is particularly decisive or omnipotent. The result was that the ocean is poorly governed, with an absence of universally agreed, binding and enforced legislation concerning how it should be managed. The ocean has fallen foul of the 'tragedy of the commons', a term first coined by the environmentalist Garrett Hardin in 1968, to denote a situation in which shared resources are poorly managed to the long-term detriment of all.

The theory of the Global Ocean Commission was that a high-level, hard-hitting international commission could change that. A series of funders got behind it, and a succession of high-level co-chairs were appointed, including David Miliband, former British Foreign Secretary and now President of the International Rescue Committee; the Former Finance Minister of South Africa, Trevor Manuel; and the former President of Costa Rica, José María Figueres. The Commission undertook over one hundred meetings around the world, published detailed reports and campaigned effectively for better policy at the national, regional and international level. The political activist and academic John Podesta joined the Commission for six months immediately prior to becoming

President Obama's chief of staff. Simon argued that President Obama's ensuing and significant ocean legacy was attributable at least in part to Podesta's exposure to the work of the Commission and his 'conversion' to the ocean cause.

Among a number of recommendations, the Commission called for the establishment of a permanent 'Global Ocean Accountability Board', akin to the Intergovernmental Panel on Climate Change, which would be responsible for providing a categorical report on the state of the world's ocean and efforts to protect it. In the absence of such an instrument, it is easy for the world to lose sight of the state of the ocean and whether it is improving (or further declining). Campaigners continue to call for such a body to be established today.

The Commissioners also lobbied for the ocean to be a stand-alone Global Goal, rather than a cross-cutting theme, and were successful when Goal 14 on the ocean was agreed in September 2015. The 'Large Ocean States' played their part in the success of this campaign, forming a dynamic coalition in the negotiations. They went on to do the same in pushing for the 1.5 °C limit in the Paris Agreement, which – if achieved – would prevent sea-level rise threatening their islands' very existence.

The Commission also advocated making massive reductions in marine plastic, which as we have seen has led to such dizzying, dystopian high concentrations of microplastics in the ocean. The Commission called for the reform or elimination of unsustainable fuel subsidies for fishermen, without which their damaging fishing practices far out at sea would not make economic sense. It also called for further implementation and improvement of the UN Convention on the Law of the Sea, which has historically lacked the teeth, and universal buy-in, which it really needs to be a success.

Simon was optimistic, in part as a result of the Commission's efforts, that public awareness about the state of the ocean had grown and would ensure the greater implementation of best practice to achieve ocean health. He hoped that there would continue to be political champions for the ocean, whether Justin Trudeau in Canada, or the Fijian government, or the Japanese who depend so much on a healthy sea. He believed that the Commission's vision of a Global Ocean Accountability Body would be fulfilled.

A year after Simon and I spoke, just before the end of 2017, the nations of the world began to take the first steps towards an international treaty to protect the high seas. In the best-case scenario, this might take two years to complete: in practice, it could be significantly longer, given the complexity of the negotiations. Nevertheless, some refer to this development as a potential Paris Agreement for the Ocean. Amongst other responsibilities, the treaty would have the remit to establish large marine protected areas where international scientists believe they should exist. It will also play its part in delivering better governance and enforcement of the high seas.

*

Path Six

My next interview was with Tony Long, another British ocean aficionado of a different kind: he had spent the best part of thirty years in the Navy, and has sailed all of the world's seas. When we met, he wondered whether he had always fully appreciated the fragility of the ocean during his career, although he certainly did now. Since leaving the Navy, Tony has spent his time leading the Pew Charitable Trusts' work to confront illegal fishing around the world. Its core focus is on strengthening and enacting a recently approved piece of international legislation, the Port State Measures Agreement, which requires all fishermen to declare their catch on arrival at a port. 55 states had signed up to the agreement by the September 2018, covering many of the key global geographies in which illegal fishing is particularly rife. A fully operational agreement would make a huge difference in preventing illegal fishing in the world, rendering it harder and harder for illegal fishermen to dock their catch in the world's ports.

Tony had a nuanced take on the different kinds of illegal fishing vessels that exist, and the scale of the damage they cause. Some of them, such as the Somali fishery in the Indian Ocean, were closely linked to pirates, mafias and organised crime, which Tony had had his fair share of experience dealing with. But others were simply poor fishermen, living in precarious conditions, including slavery or indentured labour, and often staying for months at sea. For the many millions of illegal vessels in this category, the intention is for an eventual transition to legal, sustainably managed fisheries – in which working conditions are improved, and fish stocks better managed – for the long-term benefit of the fishermen and stocks alike.

Faced with the scale of the challenge, Tony took heart from increasing consumer awareness about marine sustainability. Consumers are always shocked to learn, in exposés by the Environmental Justice Foundation and others, when the fish and prawns they buy are linked to slave labour, or the destruction of mangroves, or to an overall decline in stocks. Meanwhile, fish sourced sustainably are becoming more popular. Tony believes that properly enforced marine protected areas are also an essential solution to overfishing. He is optimistic that the world is finally beginning to tackle the lawlessness of the high seas.

I also met with a third passionate campaigner for marine protected areas, Matt Rand. His life work, based at the Pew Charitable Trusts, is to create at least 15 new international marine reserves by 2022. Since 2006, Pew has campaigned to protect Papahānaumokuākea, a large area of vitally important and biodiverse sea in north-west Hawaii. This campaign for areas such as Papahānaumokuākea has changed the paradigm, Matt argues, and now an area of the world's seas equivalent to two-thirds the size of the US has been committed to protection.

And yet, the world is still way short of the further 1,000 marine reserves estimated to be needed to ensure that the marine environment can properly function and recover. According to the science, some 30 per cent of the ocean should be protected in order to enable it to recover; by contrast, the Convention on Biological Diversity has committed the world to 10 per cent marine protected areas by 2020; and we are currently at 2 per cent. There is therefore a huge gap to fill in ensuring that some of the most important areas of the ocean are strongly protected.

Matt explained that there are great variations between these different areas, and that one had to tailor protected areas to these individual circumstances, ecosystems and geographies. He referred to a newly established Marine Protected Area surrounding the iconic Easter Island, off the coast of Chile, as an inspiring example of how to protect an area and maintain the indigenous people's ancestral and cultural rights to the sea. As a result of the protected area, the Rapa Nui people receive additional technical support from the Chilean government and international NGOs to improve their fishing practices, in return for their support in implementing and enforcing the protected area with an improved coastguard and satellite technology. Around the world, marine protected areas could be a powerful win-win for local communities as well as for their marine environment.

On the back of *Blue Planet II*, campaigners have pushed the UK for important marine areas within its own waters to be better protected. The Great British Ocean Coalition is campaigning for 1.5 million square miles of protected 'Blue Belt' around 7 of the 14 British Overseas Territories, including the Ascension Islands, South Georgia and the South Sandwich Islands in the Atlantic.

Here, there are blue whales, sperm whales, humpback whales and seabirds, as well as a huge population of penguins, rendering these islands the 'Serengeti of the sea'. The British government has agreed to increase its commitment to finance and enforce conservation there in the years ahead. Every country in the world has a role to play. And so, ultimately, does every individual citizen: for the decisions we take every day have a direct bearing on the future of this remarkable ecosystem. Now is the time to build on the unparalleled levels of awareness about the ocean to transform its prospects into the future.

Cities

As Sir David Attenborough argued when we met, the future of the planet also rests to a significant degree on how we organise our cities. In every continent of the world, amazing transformations are taking place in our urban environment. Copenhagen remains one emblematic and inspiring example of how a city can be organised sustainably from the outset. Its success is attributable in part to the 'Finger Plan' urban model it adopted in the late 1960s, ensuring that the city developed around a central organising principle, and avoiding unnecessary and unsustainable urban sprawl. The five fingers are the metropolitan train lines spread like fingers on a hand from the 'palm' of central Copenhagen, with the green spaces in between. The city has largely adhered to the plan, with even its most recent roads designed in such a way as to ensure that it is always quicker to cycle from A to B than to drive. Indeed, as much as 40–45 per cent of all work commutes in Copenhagen are made by bicycle.

To visit the city, even in the midst of winter, is to see hundreds of thousands of people cycling through the snow to work, children at the front of their parents' bikes. The Copenhagen city administration is behind the city's efforts to become a green economy, backing environmental policy and green enterprise at every step, and never losing sight of its focus on public transport. To be green is not a niche issue in Copenhagen: it is at the heart of the city's urban policy, affecting every short- and long-term decision. There are dozens of green initiatives under way in every aspect of public life, from waste management and recycling, to housing and local food markets, to ensuring that there is access to abundant green space for all segments of the population. And while the city is by no means perfect, and environmentalists there continue to fight many battles, it has undeniably set the bar high.

In London, the quixotic environmentalist and explorer Daniel Raven-Ellison is leading a campaign for the whole city to be declared the world's first National Park City. We met one sunny spring afternoon in April 2017 on a bench in St James's Park, fittingly. As much as 47 per cent of London is green space, and there are 8.4 million trees. Daniel and the campaign encourage boroughs to declare more protected areas, and to create a living network of green spaces which together render the whole city a living national park. Mayor Sadiq Khan is backing the initiative, and an active coalition of agencies and community groups is making it happen. It looks likely that 2019 will see an official declaration of a

London National Park City Partnership with representation in each of London's 33 boroughs.

London is not the only city thinking this way. A global consortium of cities – including Birmingham in the UK, San Francisco, Portland, Milwaukee, Singapore, Curridabat in Costa Rica, Wellington and Vitoria-Gasteiz in Spain – are putting a love of nature, and more diversity, wildness and green into the heart of the urban fabric. The idea is inspired by E. O. Wilson's book *Biophilia* which explores how humans have an innate love of nature and other species due to our long evolutionary history of relying upon and living among them. The cities in the Biophilic Cities Project commit to plant trees, cover roofs and walls of apartment and office blocks with mosses and ferns, and to install bird-friendly windows and roofs. They also take care to provide a haven for pollinators and other wildlife, as does our local café in Willesden Green, which provides a sanctuary for bees and makes honey on its roof.

But we have many challenges to overcome in the quest for truly sustainable 21st-century cities. There is a pressing global crisis of urban air pollution, from New Delhi to Beijing to Shanghai. Rates of air pollution are at times so severe that many people work from home to avoid the congestion and the fumes. But many more, of course, have to continue with their work in the heart of the city, often in precarious conditions, with little more than a piece of cloth or a mask wrapped around their faces. In London, one of the most polluted European cities, the air is in a bad state, and contributed to my severe asthma as a child. I worry about the impact of the quality of the air from the diesel fumes of vans, trucks and cars on Oli's little lungs. Many children growing up in the city now go to schools where the air pollution rates are as much as eight times over the World Health Organisation (WHO) standard.

I have spent days of my life sitting in traffic jams in Jakarta, Bogotá, Nairobi and São Paolo. If the air were clearer, one might be better off walking or cycling, as many in Bogotá do. In Jakarta, travel on the back of motorbike taxis, however precarious, is the quickest way to get from place to place. Faced with a particularly monumental traffic jam before a flight there once, a motorcyclist strapped my suitcase onto the back of his bike, and an hour-and-a-half's hot ride later, along back streets through shanty towns, we made it to the airport, crumpled and sweaty, somewhat worse for wear.

Cities generate a massive amount of waste and sewage, and landfill and waste systems often struggle to cope. Rates of freshwater use are often unsustainably high, while access to sufficient fresh water and sanitation – as we have already seen, and especially for the poorest and most vulnerable communities – is often highly variable and precarious. In 2018, Cape Town was under a severe freshwater rationing regime, following several years of drought: this will likely remain in place for the long term, although it has also shown how we can reduce our consumption dramatically if we have to. A number of other cities – including in Morocco, India, Spain and Iraq – face similar scarcity.

Great wealth and poverty live cheek by jowl in urban populations. Slums and shanty towns are often polluted and vulnerable to floods, landslides, water scarcity and climate change. People living on the outskirts of cities often face long commutes, as much as two to three hours each way. In Bogotá, I worked in a children's charity high up in the hills away from the city. To get there, my fellow volunteers and I took the TransMilenio, Bogotá's integrated rapid mass-transit system, to the furthermost point of the city. At the end of the line, we took a smaller bus on a bumpy trip up to the top of the hill, with the levels of poverty and environmental degradation rising as we got higher and higher. At the end of the journey, we found a very different world from downtown Bogotá: denuded hills, informal settlements, dusty streets, poor children. From such communities, many workers travel into town at the crack of dawn each day. The Trans-Milenio's long, thin red buses hurtle their way from one end of the city to the other. However, the system is now at capacity and needs major additional support, both from a revamped bus network, and from a metro which will soon begin to be built. The current Mayor of Bogotá, Enrique Peñalosa, is a celebrated advocate for sustainable urbanism based on public transport and good planning, encapsulated by his pithy saying: 'A developed country is not a place where the poor have cars. It's where the rich use public transport.' Peñalosa has been finding Bogotá a difficult place to govern, though, as a result of a strong opposition, the timelag that exists between decisions being made and real implementation on the ground, and a number of (unnecessary) battles that he has himself initiated with environmentalists and other groups. In Medellín, Colombia's second city, shanty towns are scattered in the green hills which flank the city. Here, though, a daring cable-car system allows people to commute to the centre of the city in little more than 15 minutes, soaring over their own settlements. These journeys are some of the most thrilling experiences in better urban form that I have had: at times, the cabins pass only 10 metres above the houses with their corrugated iron roofs. The cable cars bring the city together and are a universal source of pride.

Another challenge is for cities to operate more harmoniously in tandem with the rural environment which surrounds them, and on which they ultimately depend for their food and water. Cities such as New York, Bogotá and Nairobi are investing to ensure that the key freshwater ecosystems which underwrite their economies are sufficiently protected. Cities need to find ways to support better and more sustainable food production in the rural areas which feed urban agricultural markets: this often high-quality farming land is frequently being engulfed by construction, with farmers priced out of existence. As a result, some cities now produce more food in market gardens and through urban farming than they source from their immediate periphery. And while urban farming has its place – and indeed can be shown to be highly productive, as in the case of Cuba, where Havana and other cities as a result of the US's economic sanctions produce a major percentage of the city's food – it can never wholly replace the role of food production in rural areas. In summary, the relationship between the cities and their rural environment can either be beneficial to the climate and the natural world, or destructive: it is our job to ensure that it is the former.

Many cities are acting, with vision and commitment, to become rapidly more sustainable. Here, they are blessed with a big advantage: they can act quickly and decisively to implement exciting and innovative new ways of addressing environmental challenges such as pollution, waste and deficient public transport – often much quicker than national governments. Many cities, and their mayors, belong to coalitions in which they share the best ideas, such as the Global Covenant of Mayors for Climate and Energy, which brings together several thousand mayors from around the world; the C40 Cities Climate Leadership Group, a high-level coalition of the world's largest and most active cities on climate change; and the Local Governments for Sustainability (ICLEI), which is particularly strong on climate change and biodiversity conservation in cities. These groups provide an opportunity for cities to learn from one another, and replicate successful initiatives.

In a hectic and often heavily polluted urban environment, the human need for nature is only heightened, with natural spaces providing a vital oasis for children amid the stresses and strains of contemporary urban life. One particularly critical priority is to provide more space for nature in our cities, with more trees and parks, and greater environmental education.

To understand the issues better, I went to interview Mark Watts, Executive Director of the C40 network, which counts amongst its members London, New York, Shanghai, São Paolo, New Delhi and Mexico City. We met in the bustling office of the C40 in Central London. Mark brims with enthusiasm and insights from cities around the world – informed by his early experience working in the mayoral administration of Ken Livingstone and his subsequent travels on behalf of the C40.

When we met, the Mayor of Paris, Anne Hidalgo, was chairing the C40. Hidalgo had been making real strides during her tenure, albeit not without controversy, to make Paris more environmentally minded. She pledged to ban all diesel vehicles by 2020, to pedestrianise one of the banks of the Seine, to ban disposable plastic and to support start-ups committed to solving particular environmental problems. Mark was particularly enthused by Hidalgo's scheme to offer communities across the city the chance to support locally conceived re-greening projects. Her fund has already supported 40 urban regeneration initiatives. Where once there had been a car park or a disused public space, the mayoralty had provided funding for them to be converted into green roofs or community gardens.

In Portland, Oregon, Mayor Charlie Hales had blocked

further expansion of the city's oil refineries and instead focused the city's finances on an expansion of solar energy and jobs, arguing – correctly – that more people would stand to benefit in the long run. The city has been undergoing a significant green renaissance, from which other cities in the US have sought to learn.

Curitiba in Brazil is known for its participatory budget and planning processes, and its long-standing commitment to sustainable urban planning which dates back to the 1970s. The city has completely reformed its public transport system and dramatically reduced its waste. Sydney's Lord Mayor, Clover Moore, has also enacted ambitious urban reforms, retrofitting buildings to be more energy efficient, even when in national politics there has been much climate scepticism and reneging on environmental policy.

Although less well known and understood in the West, some of the biggest changes are happening in Chinese cities: Beijing, of course, but also Shenzhen and Nanjing, where electric vehicles are being introduced in record time, and where 4,000 or so electric buses have been ordered, more than the whole of Europe combined.

Mark spoke passionately about how the C40 and the Global Covenant of Mayors herald a new form of politics. For while cities and coalitions cannot displace the nation state, it is significantly easier for city leaders to collaborate with one another and share ideas and good policies. And so it is that cities are both a major driver of economic growth and well-being, on the one hand, while also holding the promise of a new and better model of global governance.

Powerful proof of how important enlightened cities can be came during the writing of this book. Mark and I met before Trump stated he would pull the US out of the Paris Agreement. Once he announced his intention, many of America's foremost cities rallied, strengthening the commitments they had already made. A number of coalitions – including the US Climate Alliance, 'We Are Still In' coalition and America's Pledge – have been established. Several hundred non-federal actors in the US – including cities, but also companies, Native American tribes, NGOs and whole states – have joined these coalitions to support the Paris Agreement. Combined, these supporting cities and states now account for nearly half of the US population and more than half of the US economy. If they were a country, they would be the third largest in terms of GDP and the fourth largest in greenhouse gas emissions. Domestically, these movements are working to dramatically reduce greenhouse gas emissions. At the global level, the coalitions are also engaging diplomatically and maintaining a credible US presence in the climate change negotiations. Together, they have enabled observers to feel that the US is still committed to climate action, even if its president and his team are not.

Mark's top priority for cities is to achieve better mobility, so that citizens can benefit from what cities do best: allow people 'to connect easily with one another'. The challenge is to build compact, dense cities in which it is possible to move around, get to work and access amenities quickly without a car. The world has learned that the car-based sprawling model of urbanism no longer works; travel by cycling and walking provide efficient cheap mass transit, as well as a low carbon city. It is much better for people's health, too.

Only 15 years ago, as a transport adviser to Mayor of London Ken Livingstone, Mark gave a speech at an urban planning conference arguing that the future of London would be built around cycling and buses; people laughed. The view then was that trains, roads and the underground were the future; bikes had to be eliminated or firmly discouraged. Now, according to Mark, 'nothing short of a revolution' has taken place and London is replete with cyclists and cycle lanes, for which the former Mayor Boris Johnson also deserves some credit. The current Mayor Sadiq Khan used some of the Olympic investments to encourage a further shift to buses and bicycles, and Mark is pleased that air quality is top of Sadiq's agenda. An active civilian movement led by the London Cycling Campaign is pushing strongly for further improvements to London's cycling infrastructure.

If every mayor were to adopt proper road pricing, this would unlock a transport revolution, as cities as diverse as London, Singapore and Stockholm have shown. In the late 1990s, when Mayor Ken Livingstone first made the argument for a congestion charge, many Londoners complained. But he was backed up by the central government of the time, in particular by Prime Minister Tony Blair and Deputy Prime Minister John Prescott, and Londoners soon noticed the difference: buses gained their own lanes and became much quicker and more predictable. The congestion charge is now an integral part of modern London. Sadiq Khan is now expanding the principle of the charge to encompass diesel pollution, which is responsible for the bulk of the air pollution crisis facing the city.

To achieve such a shift in New Delhi or Beijing is not straightforward, but it is almost impossible to contemplate without an effort to reform fuel pricing, to regulate taxis, to re-designate road space for clean mass transit and to make better plans for public transport. This requires brave political leadership from the mayor and national government to take on the lobbies representing the vehicle sectors. Across the world, including in New York, well-conceived schemes can flounder in the absence of government support, as both Mayor Mike Bloomberg and his successor Bill de Blasio learnt with recent attempts to implement their low carbon transport plans.

Progress elsewhere is encouraging. Knowledge of the impacts of air pollution has led to popular uprisings and increased citizen activism, a potent combination. Many efforts are under way to equip city dwellers with up-to-date information on the air quality in their cities, with which to hold their mayors and national leaders to account. Cities are also becoming more thoughtful about the food they eat, with a growing market for sustainable and healthy food. In Bogotá, there is a vibrant market in Guasca, just outside the city, which has committed itself to 100 per cent organic agriculture. Each weekend, thousands of *bogotanos* descend into the *sabana* – the verdant landscapes surrounding the city – to stock up on their vegetables and meat for the week.

As we have seen from the London National Park City campaign, educated urban populations are also clamouring for more access to green space and for more trees. New networks of food and urban agriculture are springing up in many cities, with efforts to make schools and homes for the elderly linked to locally produced, organic food, and to provide schools and prisons with the opportunity to grow more food themselves. In China, urban agriculture in some new cities which have been built on what was previously agricultural land are now producing a greater volume of food through urban agriculture than the equivalent amount from the rural area before.

Cities are also making huge progress in addressing their waste, which manifests itself differently in cities of the north and south. In much of the developing world, most waste is still not being fully collected, informal dumping is rife and causes significant greenhouse gas emissions: around 25 per cent of Latin America's overall emissions are from methane from the continent's big city dumps (whereas in the West this figure is in the range of 2–5 per cent). The challenge in many developing countries is to manage waste in a responsible way. In San Francisco, by contrast, the city is close to achieving zero waste going to landfill: the city has embarked on an effort to establish a circular economy which eliminates the concept of waste and which reuses and recycles the vast bulk of material it generates.

A number of the Scandinavian and North American cities own their own energy utilities, and have enabled major new developments to be equipped with on-site renewable energy. Across the townships in Johannesburg, including Soweto, solar energy is providing hot water and lighting health clinics and schools. But cities will need to be given more power to enable this energy revolution to take place, as some of the key decisions on energy infrastructure are still often taken at the national or state level.

On the question of housing and infrastructure, the aspiration is to build compact and dense cities, rather than sprawling ones. Creative architects and developers are coming up with ways to build high-quality living environments within compact areas. Urban planning, long felt to be unfashionable, has come back into vogue. There is also renewed interest in building well-designed, mid-sized, compact apartment blocks rather than the tower blocks of old.

In Kuala Lumpur, in early 2018, the World Urban Forum met to broker an international agreement on how the cities of the future should be. With many exceptions, the view is that cities are not yet on track: much more will need to be done. But the world has a very clear vision of what this urban future might look like, and it is ours for the making.

Waste

The next, pressing way of saving the world is to develop a circular economy, in which we recycle goods back into the mainstream and eliminate waste. In India, the company Banyan Recycling has developed one of the world's first plastic recycling companies, assisting global brands to use more recycled plastic in their operations. The company has developed a plastic cleaning technology which converts post-consumer and post-industrial plastic waste into high-quality recycled granules which are comparable in performance and quality to virgin plastic. Banyan has established a data intelligence platform that links thousands of informal recyclers with their supply chain; the platform also works with cities to reduce their waste and to manage it more effectively. The cleaning technology removes coatings, inks and contaminants by using more environmentally friendly detergents and solvents. Banyan works with one of India's leading car companies to make new bumpers from discarded ones, and with a global cosmetics company to make new bottles from old. The company pays its staff a decent wage, as well as a pension and health insurance, and has won a number of awards for its environmental performance. It provides an excellent example of the almost infinite possibilities for a circular economy to take flight.

Elsewhere, eleven global companies – including Unilever, Walmart, Evian, Mars and Ecover – recently announced that they will ensure all their packaging is reused, recycled or composted by the year 2025. These companies are together responsible for about 6 million tons of plastic packaging. By signing up, they have shown that it is possible to go beyond the so-called 'take–make–dispose' model of consumption and instead seek a circular economy which causes no waste. And while each of these companies, as well as thousands of others, could do very much more to reduce the quantities of plastic in use in the world, it is a start.

Closer to home, the British department store John Lewis is piloting a circular economy incentive scheme, pledging to buy back old clothing people no longer want, at a modest price, and ensuring that those clothes are recycled rather than sent to landfill. The service is enabled by the start-up company 'Stuffstr', whose motto is: 'Don't trash it. Recirculate it.' In our own, modest bid to lower our household footprint, Davina and I made use of another wonderful start up – 'Reyouzable' – which delivers rice, beans, washing-up liquid, detergent and a host of other good things to our flat in London without the use of any plastic packaging. The founder, Gavin Prentice, did the deliveries himself, arriving by public transport with muesli oats delicately wrapped in paper bags for subsequent decanting – although he has recently decided to give it a rest, noting the widespread take-up of his idea by other companies. It is possible to live with very little waste if you really put your mind to it, but it requires constant discipline and effort.

This is just as well, because humanity is still producing a colossal, unnecessary and frankly 'wicked' – as Sir David Attenborough referred to it – amount of waste. The world wastes a third of all the food we produce. Imagine how many millions of mouths we could feed, and how much pressure we would take off the world's ecosystems, if we ate all that we produced, rather than letting it languish either at the farm or in the refrigerator.

The machines that we build do not last, nor much else of what we buy. We are plastic bag, bottle and straw addicts, with so much of this material ending up in landfill or in the ocean. But our landfill sites are fast filling up, as Paul Polman, CEO of Unilever, told me. Our clothes – often made by workers in appalling conditions – and fashion trends don't last very long. We waste huge amounts of energy, as any walk through London late at night will tell you, its empty office blocks largely lit.

It does not have to be this way, and it is eminently possible to establish an energy efficient and circular economy, in which products are used for as long as possible, with the maximum value extracted from them, and products and materials recovered at the end of their life. Around the world, countries, companies and societies are finding ways to reduce and recycle waste as well as to create the kinds of circular economies described above.

The United Kingdom achieved a 21 per cent reduction in food waste from 2007 to 2012, thanks to effective advocacy campaigns and a culture of thrift brought about by the financial crisis. Companies have made significant pledges to reduce food waste in supermarkets and across their supply chain. A grassroots movement in Britain has found willing partners across much of the world, from Argentina to Australia, Nigeria to the US.

I went to interview one of the most prominent campaigners on food waste, Tristram Stuart, to find out more. At the offices of his campaigning organisation Feedback, he began by saying, 'Food waste is in and of itself a colossal scandal, as well as an equally colossal opportunity. If we need to increase food availability for a population of 9.8 billion by 2050, while at the same time reducing the environmental impact of food production, food waste is a low-hanging fruit: we can eat and enjoy our food rather than throw it away.'

Tristram re-articulated the fundamental paradox of our food system: that we simultaneously have a public health crisis in over-consumption, while at the same time a billion people are malnourished. The world's agriculture and food system, responsible for the biggest impact we're having on the planet through deforestation, freshwater use, soil erosion and biodiversity loss, as well as being the biggest source of CO_2 emissions, can instead be turned 'into the most powerful tool for putting carbon in the soil and out of the atmosphere, creating habitat and food for everyone'. By tackling something egregious, addressing food waste also becomes something joyful and celebratory.

In the UK, food waste begins right at the level of the farm, as a result of the 'ridiculous purchasing policies of the big buyers,' Tristram said. The supermarkets, five of which account for 80 per cent of the country's groceries, wield massive power in terms of who they do business with. As a result of their 'totally unnecessary cosmetic standards', as well as price fluctuations, orders are often cancelled and farmers have to incur the cost, both financial and environmental, when fields full of carrots and onions are wasted. Cosmetic standards are particularly unfair, Tristram continued, given that nature 'produces diversely not uniformly' – food is not meant to look uniform, but instead healthy carrots and potatoes come in all shapes and sizes. We are not only wasting fruit and vegetables, but also lots of meat 'through offcuts and ugly bits of animals which have fallen out of favour'. In our already over-exploited ocean, some species of perfectly edible fish are also being discarded for similar illogical reasons. Meanwhile, 'supermarkets pile their shelves high in order to convey a sense of Cornucopian abundance' for marketing purposes which influence us to buy more than we could possibly consume. 'We have evolved,' Tristram said, 'over the past two million years in an environment characterised by scarcity, by and large, in which we took what we could and hoarded what we could.' The result, in rich countries at least, is a world of permanent abundance in which we waste 20 per cent of what we buy.

In Senegal, by contrast, the situation is different: a huge number of mangoes are grown to be exported, but 'an awful lot is wasted'. They rot on the trees or, due to a lack of agricultural infrastructure and processes, are not preserved. Food is often kept in sub-optimal conditions in which vermin or mould can spoil it. Local people are sometimes forced to consume this spoilt food for lack of an alternative, leading to cancer, stunting and hepatitis in children. 'The challenge here is to invest in agricultural infrastructure and processes that ensure African farmers, and farmers elsewhere, can make best use of the products they have grown,' Tristram says, with this investment serving as a route out of poverty and a means of increasing food availability. Tristram argues that we must also reconnect with our food production: 'It is almost impossible when in a supermarket to bring to mind the resources, the effort, the people, the love even, that went into that food, and so we treat it as a disposable commodity.' Instead, we must make better choices 'for health, for fairness, for equity, for the economy, for our own pockets'. It is a question of making 'individual and collective changes in our values and choices; revising and rewiring our habits, in order to rewire the food system'.

Feedback runs a series of 'Feeding the 5000' events, in which they provide thousands of free and delicious meals made of discarded food, which would otherwise have gone to waste, to people in Trafalgar Square, New York and elsewhere across the world. This is a symbolic way of demonstrating to the public the positive outcomes that would follow from addressing food waste.

In 2001, when Tristram began campaigning, working on food waste was a fairly lonely place: now, its reduction is a global movement and a booming industry. Wherever you care to look, there is action under way: the month after we met, he went to Argentina, where there is a growing food waste movement, and the Mayor of Buenos Aires, Horacio Larreta, had organised a big public event in the city. The US government is acting on the issue; many of the world's largest multinationals now have measurable commitments to reduce food loss and waste in their supply chains; and many national governments are now enacting legislation to address it. Significant reductions have occurred. Global Goal 12.3 – a sub-set of Global Goal 12 on sustainable production and consumption – is to halve food waste by 2030. A real culture shift of mass behaviour change, and corporate policy change, is under way, led from the grassroots.

Tesco is a case in point: after 'Horsegate', a 2013 scandal in Europe in which meat sold as beef was discovered to be horsemeat, Tesco lost €360 million in value, 'was on its knees' and called in Feedback for advice on how to address the outcry. Feedback encouraged Tesco to be transparent: they became the first supermarket to report on food loss and waste in their supply chain, making 'significant and far-reaching reductions across the UK'. The supermarket companies are now on the run, with a number of them demonstrating real leadership and genuine engagement, and the laggards trying to work out what to do to catch up. Brand damage from examples of bad practice is significant: consumers will not settle for Kenyan or Costa Rican farmers producing their coffee beans, being paid less than $2; or for farmers having their orders cancelled at the last minute; or for being forced to trim 20 per cent off their beans so that they would fit in a 9-centimetre punnet. Tesco is now one of the champions of global food-waste action bringing together businesses, governments and individual citizens demanding change. For its CEO, Dave Lewis, tackling food loss and waste is fundamentally a moral issue: for someone who has worked a lot in Africa, he thinks it is fundamentally unacceptable.

Tristram's new initiative, 'Toast Ale', began after an encounter with a brewer in Brussels – the Brussels Beer Project – who had had the fantastic idea of turning 'waste bread into craft ale'. 'Bakeries routinely overproduce and overstock', as Tristram's book *Waste* described, with sandwich manufacturers for example throwing away the crusts of all their loaves. The brewer came up with a recipe based around discarded bread. The very first bottles, sampled live on a Jamie Oliver TV programme, were judged to be 'blooming delicious'. When we met, Toast Ale was flourishing, with franchises soon to open in Iceland, South Africa and the US, and profits going to Feedback. This was in Tristram's words a way of 'drinking our way out of a global problem'.

I wondered whether Tristram, who has written an

intellectual history of vegetarianism, felt that global vegetarianism was the answer to the world's environmental predicament. His belief was 'that the more vegetarians and vegans the better', but that a dramatic reduction in meat and dairy consumption is more effective than absolute abstention.

Food waste is one important part of the story. But a fully circular economy encompasses every other area of human activity. The quest is to establish an economy which is regenerative by design: from the outset, an economy which is built around principles of regeneration and reuse. Another way of describing the circular economy is 'cradle-to-cradle', where products and materials are designed and used so as not to lose their value but to be endlessly reused and recycled.

Martin Stuchtey is Founding Director of the 'B Corp', or 'company with a purpose', Systemiq, as well as a Professor at the University of Innsbruck, and a clear-sighted thinker on the circular economy. According to him, an average human being today consumes nearly 11 tons of resources from the earth per year, whether in the form of metals, wood, food or nutrients. These materials sustain the world's consumption, including our energy, vehicles, mobile phones, buildings, clothing, food and transport. But whereas in nature nutrients circulate in 'safe, natural metabolisms', with waste becoming food, and materials never ceasing to be nutrients, in the linear industrial system most materials are simply used once, and altered in ways which render them useless, or at least nearly valueless, to the next generation of users.

But in Martin's view regenerative, high-productivity systems can become the norm rather than the exception. The goal should be for materials to be reused many times and to create a market for these materials much bigger than the one which exists today. In the future, it should be seen as a competitive advantage to have a strong circular economy and a bank of these reused technical materials. The circular economy then ensures a much better and more sustainable use of finite natural resources, such as timber from forests, much lower than today's excessive rates of extraction and use.

At the Systemiq office, overlooking St Paul's Cathedral, Martin explained that the circular economy ultimately implied the search for a new way of doing things. This is 'not an incremental change, but rather a fundamental rethinking of the systems we use, 200 years on from the birth of the industrial revolution'. This new economy will require a new way of doing business: hence, the company he and his partners had established. The world's concept of growth must be decoupled from resource use; technological systems exist to do this. This is 'no longer a utopian dream'.

Even plastic, that seemingly most permanent of all wastes, can be derived from renewable sources, either biologically or from other plastic waste, Martin said. Of all the plastic the world produces, 34 per cent currently ends up in its ecosystems, the bulk of it in the ocean. To combat this, we need to create a market for plastic such as the one we have for aluminium, gold, silver and paper, where none goes unused. The technology exists to do this at scale, with plastic well suited for repurposing and recycling. In such a system, Martin imagines, 'One would then have the health and hygiene benefits of plastic as a material without the massive costs.' Plastic is the litmus test for the move from a linear to a circular economy: 'If we can do it with plastic, we can do it with anything.' The subject has generated a massive amount of momentum, with CEOs from the major consumer goods companies addressing the issue, knowing that if they do not, 'someone else will write the rules', and wield a huge amount of power in this emerging market. We are still, however, in the 'headless chicken phase' of dealing with the issue. By comparison with the energy transition – where all decisions, however decentralised, ultimately lead to emissions reductions – plastic is a vexing problem. An attempt to deliver better plastic, recycling or a substitute to plastic in one area might be the utterly wrong answer in the overall system. This is a 'team sport': the solutions chosen in one area need to fit in with what we do everywhere else.

Individual countries are also taking action, such as in Indonesia, where the government has committed $1 billion to address the plastics crisis. The Indonesian plan on marine debris includes legislation that extends responsibility for plastics to producers, as well as providing incentives and resources to assist the industry to change its practices. And yet 'the government finds it difficult to assert these rules', Martin believes, and the issue continues to be acute. No one has as yet found a model that works for communities and businesses at the same time.

Martin is also a passionate farmer in the Alps, where he has improved the quality of his family's land and the levels of biodiversity on the farm by 'thinking in circles'. In his view, a well-managed farm is the ultimate example of a circular economy. He has realised that the better he farms, the fewer inputs he needs to add. And by tailoring the agricultural goods he produces to the demands of individual consumers, he is creating a direct market between producer and consumer – taking out the intermediaries, to the benefit of farmers and consumers alike. He foresees more agricultural produce coming from peri-urban (between town and country) areas in the future, leading to a more circular system of nutrient flows between cities and their rural surroundings.

The internet will also play a role in ushering a more circular economy into being, by helping us to match every component, part and ingredient from one project or consumer to the next and keeping levels of value high across many life cycles. In a highly connected world of almost 10 billion people by 2050, in which we threaten to overstretch resource boundaries, a circular economy is our best bet. Smart metering, autonomous cars, better management of space in office buildings, industrial protection, 3D printing and distributed manufacturing will all contribute to our cities becoming more inhabitable and productive. Martin concluded: 'All of the different elements and pieces of the puzzle are out there, pointing to a circular company, industry and even country – but we are not there yet.'

The British sailor Ellen MacArthur, who broke the world record for solo sailing in 2005, described her experience on the expedition in the *Guardian*: 'I remember quite poignantly writing in the log on the boat; "What I have got on the boat is everything." It really struck me that you save everything, everything you have, because you know it's finite, you know there isn't any more. What you have on that boat is it, your entire world.'

After the expedition, MacArthur spent a fortnight on an island in the Southern Ocean making a film about the albatross. 'It gave me time to reflect and it made me think even more deeply about resources,' she said. 'You see the empty whaling stations down there and you real-ise that was just a resource – they pulled out 175,000 of them . . . and then there weren't any to pull out.' Ellen has subsequently become one of the most articulate and convincing exponents of the importance of a circular economy.

The world's textile industry is her latest target. Every second, the equivalent of one dustbin lorry of textiles is burned or dumped in landfill. And every year, an esti-mated $500 billion of value is lost due to clothing that is barely worn and rarely recycled. On current projections, by 2050 the fashion industry will use up a quarter of the world's carbon budget. At least half of the fast-fashion purchases people make are disposed of within a year, most typically in landfill. The clothing industry is also responsible for releasing half a million tons of micro-fibres into the ocean every year, which equates to 50 billion plastic bottles. Microfibres are impossible to clean up and can enter the food chain. MacArthur argues persuasively that there needs to be 'a new fashion industry for the twenty-first century, in which clothes are made to last longer and to release fewer toxins into the environment'. She also says that society should explore ways in which clothes can be rented and shared, as well as made of less destructive materials. To achieve such a transformation will require social change, new materials and better technology; it can certainly be done.

Two further examples of the circular economy, among many, strike me as inspiring. The first is the social enterprise GlobeChain, set up by the young British woman of Iraqi heritage, May Al-Karooni. She was upset to learn that when moving office in the city from one side of the street to a block five minutes away, the entirety of her office's carpets, desks and computers were to be scrapped, and the company would source new ones. She set up GlobeChain to address this pointless waste. Using an on-line platform, her company now assists social enterprises and charities in laying claim to office-related materials which would otherwise go to waste. The company has flourished, matching huge amounts, from National Health Service equipment and office stock to furniture; and the appetite for the company's services seems limit-less, with some 10,000 members and counting. A number of spin-offs have taken off around the world.

The second is in Ghana: a local social enterprise with Dutch backing, Safi Sana, has found a way of turning organic and faecal waste into electricity, soil conditioner and irrigation water. The model is taking off worldwide. In the settlement of Ashaiman, home to 250,000 people, Safi Sana is collecting waste from improved public toilets. The organic waste is being used to produce electricity, biogas, organic fertiliser and seedlings. The cost of maintaining the factory and its operations is covered by the sale of the products, ensuring that this is a long-term, financially sustainable model. The staff are almost all local, and benefit from an intensive training programme. The resulting business takes pressure off fresh water and enables a circular economy in that most fundamental of all waste products, human sewage. Here too, there is always hope.

People

All of my ten paths to a more hopeful future are profoundly to do with people and their lives, well-being, health and dignity. There is no dichotomy between a concern for the environment on the one hand, and humanity on other. The Global Goals are an eloquent manifestation of how the two are intricately and inextricably linked. There is no good life on an irreversibly depleted planet. But nor can there be environmental harmony and restoration without human well-being at its heart. The abiding task of our times is to ensure a good life for all, within our environmental boundaries and constraints.

One man who understands this relationship in depth is Dr David Nabarro, a British doctor with a long-standing career in international public health, the British civil service and the UN system. In 2018, David won the World Food Prize for his achievements in the fields of nutrition, global food security and childhood stunting. Among many plaudits, David led the global response to the Ebola crisis at the UN, and when we met, described its beginnings in the interior of Guinea, the initially incorrect diagnosis and its subsequent rapid spread in towns, with the numbers of cases doubling every three weeks. When David came into the post, he immediately requested a twenty-fold increase in response capacity from the UN Secretary-General and, working closely with ministers of health from the affected countries, pulled together a successful global coalition to respond to the crisis. The curve of the disease started to bend and plateau by the end of 2014 and a steady decline in cases continued through 2015. While the world must remain vigilant, as the recent outbreak in the Democratic Republic of Congo shows, the global response to the Ebola crisis was one of the more impressive feats in global public health and the multilateral system in recent years.

David is an enthusiastic advocate of the Global Goals, in whose negotiations he was actively involved as a Special Adviser to the UN Secretary-General. In his view, Goal 13 on climate action is the most important: 'the critical path around which everything else hangs'. He is worried by the health implications of climate change, from extreme heat to new patterns of disease spreading, including malaria and dengue. But he is optimistic about the human capacity to solve problems. His professional experience, including as a doctor, in over 50 countries has taught him: 'People have the capacity to ensure the world is a sustainable and renewed place.' David's profound humanism and commitment to purposeful and collaborative international action is the spirit which informs this book.

There are three core sets of challenges when it comes to people and their relationship to the environment: the whole question of population growth, size and demographic change; what we all eat; and how much we all consume. The three issues need to be understood together rather than separately.

To tackle population first: the UN predicts a global population of 9.8 billion by 2050 and 11.2 billion in 2100. Sir David Attenborough is surely right to express a concern about the collective impact of this number of people on the natural environment and on our greenhouse gas emissions. Many of the targets the world has set itself – on food security, water, climate, cities – are already daunting. They will become much more difficult to achieve in a world of an additional 2.6 billion people by 2050, the majority of whom will be living in developing and low-income countries, where pressures on finite natural resources are already high.

But to change human behaviour in this fundamental area of human experience is not straightforward, as the UN expert on girls' education, Nora Fyles, expressed: 'The drive for sex and pleasure, and the security of children to carry on the family line, food for self and family, are among the most difficult behaviours to influence.' 'As a parent,' she said, 'you know better than anyone the centrality of what is on the plate, the importance of hugs and security of family, the allure of one more or a bigger object or toy, and the challenge of teaching and learning about the responsibility to put away one's toys and rubbish.' As a global community, we face 'the monumental task of relearning how to be people on the planet'.

The answer is to provide access and ensure people's rights to reproductive health care around the world, one of the most important causes of our time, irrespective of the climate and environment: a matter of human rights and dignity, first and foremost, and above all, for women and girls. All of them should have access to the education and reproductive health care they need to live fulfilling lives. 'And at the right time,' Nora added: 'girls of 12 or 13 are considered old enough to be married in some countries, but are generally not considered old enough to have access to the information they need to either protect themselves or make informed decisions.' The overarching problem is 'girls having unprotected sex and having babies as children themselves', then going on 'to have families of seven or more before becoming grandmothers at age 30'.

To illustrate the point, it is worth looking at the world's ten fastest growing national populations and what they're projected to look like by 2050. India, which today stands at 1.2 billion people, is projected to have a further 467 million people by 2050. In Nigeria, today's population of 180 million will be nearly 300 million by 2050. Pakistan will go from 193 to 294 million; Tanzania, from 55 to 148 million; the US from 327 to 420 million; the Democratic Republic of Congo, from 79 to 161 million; Ethiopia, from 104 to 168 million;

the Philippines, from 103 to 165 million; Uganda, from 41 to 102 million; and Kenya, from 48 to 104 million. These are big numbers, especially when one considers that in India – by way of one example – 600 million people are already at risk of water stress today. Surely the additional 467 million people, all being equal, will place further strain on what is clearly already a scarce and over-stretched resource. This is not to be Malthusian, and to predict that calamity awaits; it is merely to express a logical concern.

It takes a generation at least to reach a stable population: for example, South Korea now has a fertility rate – the average number of children born per woman – of approximately 1.3, having introduced a fairly strict family-planning regime as far back as the 1970s. Its population is still growing, though very slowly. Many large countries – including India, Indonesia, Bangladesh and Vietnam – now have so-called replacement rates of childbirth: i.e. 2.1 or 2.2 (with Latin American countries lower than this). But because there are so many women of childbearing age, their populations will carry on increasing for some time to come. It is interesting to study which factors other than a 'strict family-planning regime' could lead to the slowing of population growth. In each of these countries, a major focus on education for girls as well as boys, and labour markets open to women as well as men, giving them access to real opportunities, are important contributing factors. Rapid economic growth has gone hand in hand with population stabilisation, so poverty is another factor. This is not a simple story, but a complex dynamic where gender relations and women's rights, education, labour market structures and economic growth all play a role.

Half the population growth of the next thirty years will take place in Africa, which anticipates a further 1.3 billion people in the continent by 2050. This will occur despite a likely reduction in the fertility rate from 4.7 births per woman between 2010 and 2015 to 3.1 births between 2045 and 2050. Africa is therefore a particular priority in ensuring girls' education through secondary school, access to health care and reproductive health care across the continent. But access to the last must be voluntary. Negative social and gender norms, and the institutions which are an extension of these norms, reinforce men's right to have sex, and disallow girls and, in some cases, women to have access to reproductive health information and services. When the President of Tanzania, John Magufuli, stated publicly in 2017 that no girl who has been pregnant will be permitted to return to school, this was a breach of their human right as expressed in Global Goal 4. When girls are not permitted to have access to information about reproductive health, this is also a breach of their human right as expressed in Global Goal 5. Even today, for hundreds of millions of women around the world, access to safe and affordable sexual health care remains out of reach.

If the global effort to reduce fertility rates through intelligent policy and better health care is stepped up in such that a way that we reach a global population of nearer 9 billion, rather than 9.8 or even 10 billion by

2050, and nearer 10 billion rather than over 11 billion by the end of the century, this will be of significant benefit to the environment. For while there are great variations between the scale of the impact different individuals have on the environment, every human life has an environmental footprint. And so the fewer of us that there are, the lighter our collective impact on the natural world.

And so what needs to be done? Policymakers should begin by making the link between education and climate change, not just in theory but in funding decisions and programming allocations. The Global Goals are intended to bring historically disparate goals together: this is evidently the case on the issues of education and climate. Undertaking a big global effort to ensure universal secondary schooling for girls would bring the two together. Often, the cheapest, most cost-effective mechanism for reducing emissions does not seem to have been considered by the international community: many argue that it is education, or more specifically girls' education, that is as likely to result in lower carbon emissions as any of the other shifts described in this book.

The impact of education is enormous. Better educated women have far fewer children than women with little education. The difference between the high and low population growth assumption is 2 billion people by 2045 and over 5 billion people by 2100. Almost all of this difference depends on the assumption made about fertility. In Africa, improved education could result in 1.8 billion people fewer than the UN median variant suggests is likely by 2100. If education could be scaled up faster, the impact would be even more dramatic. It is one of the most important things we can do, as well as being deeply right.

The second means of lightening humanity's impact on the planet is, as we have seen, to reform the global diet, and indeed the whole food system which underpins it. To a significant degree, this is about much of the world's population eating much less meat, in particular red meat, and increasing their consumption of a 'plant forward' diet based on fruits, vegetables and grains. But it is also about ensuring that the 815 million malnourished people – 1 in 9 of the world's population – who go to bed hungry every night have access to the food that they need to live healthy and rewarding lives, meat or no meat. Chronic malnutrition in the early years of life leads to stunting, preventing children's brains and bodies from growing properly. This damage is irreversible and has significant consequences for their adult lives, education and employment prospects. It is also morally untenable that we should be in this situation.

In addition, the world needs to tackle the growing epidemic of diabetes, overweight and obesity, through better food policy, encouraging healthier diets and exercise, and putting progressive sugar and anti-junk food regulations and laws into place. Global obesity rates have effectively doubled since 1980, with 30 per cent of the world's population now overweight or obese. This equates to 2 billion people globally, including 124 million young people. Many countries are now affected by the paradoxical 'double burden' of growing rates of obesity and simultaneously high levels of malnutrition and child stunting. The CEO of Unliever, Paul Polman, is right when he says that a Martian visiting earth would be shocked by how we have organised our food system to deliver such manifestly perverse outcomes.

Many countries are beginning to take dramatic and far-sighted action to address these worrying trends. China – which produces and consumes 28 per cent of the world's meat, and which has over 100 million people with diabetes (some 10 per cent of its population) – has pledged to reduce its meat consumption by 50 per cent by 2030 in order to reduce its climate footprint and address some of the country's major health challenges: obesity, cancer, diabetes and heart disease. The Chinese Ministry of Health guidelines recommend that the population should consume between 40 and 75 grams of meat per person each day. In 1982, the average Chinese person ate 13 kilogrammes of meat per year and beef was known as 'millionaire's meat' due to its scarcity; today, the average is 63 kg of meat a year, with this anticipated to grow by a further 30 kg by 2030. Chinese celebrities have been enlisted in public service announcements aimed at slowing the growth of meat consumption in their cities.

In the West, pioneering work is under way to encourage restaurants and supermarkets to shift their marketing, menus and cooking practices to favour more vegetarian or 'plant-based' dishes, without using the 'v' word. Very few of the popular options in a restaurant or in the supermarket in the US or UK are currently vegetable-based. And yet consumers, presented with a 'superfood bowl', or the Middle Eastern dish *shakshukha* (baked vegetables and eggs in a cast-iron dish), are more likely to choose these over traditional options. A 'nudge' based on sound principles of behavioural economics and human psychology helps people in choosing options which are fundamentally much better for them. And so it is that diets might shift quickly in our lifetimes, buoyed on by a growing awareness of the importance of healthy food and a good diet – as much a phenomenon in modern-day India, South Africa or Brazil as it is in Europe and North America.

To take the US alone, Americans eat approximately 10 billion burgers each year. If we were to replace 30 per cent of the beef in those burgers with mushrooms, this would reduce greenhouse gases by the equivalent of taking 2.3 million cars off the road. It would also reduce demand for irrigation water by 83 billion gallons per year, and demand for global agricultural land by more than 14,000 square miles. This is also a business opportunity: sales for the US-based company Sodexo of their beef-mushroom blend have been going through the roof. In 2017, the national burger chain Sonic committed to trialling 'Sonic Slingers', a blended burger, across the country: 'its juiciest burger ever'. The sky is the limit for these kinds of changes, which are happening at a precipitous rate, as is the development of artificial, lab-grown meat.

The third, related challenge is to reduce consumption.

Much – if not most – of what so many of us consume we do not really need. And yet the environmental consequences of this level of consumption are significant. One new polyester shirt, on its first wash, will leak several thousand damaging microfibres into the ocean. Christmas decorations and supermarket packaging lead to huge amounts of wasted material every year, as Sir David Attenborough lamented to me when we met soon after Christmas. The economist Tim Jackson puts it well in a celebrated quote: 'We spend money we don't have, on things we don't need, to make impressions that don't matter.' We live increasingly in a culture defined by shopping and materialism. People love 'stuff' and gain pleasure from the frequent purchase of unnecessary items. Many of the things we buy have a short shelf life: some are made deliberately with 'inbuilt obsolescence', a ploy by manufacturers to ensure that they can replace their products in a few years' time. Our mobile phones – which we automatically upgrade every few years – are built with materials such as coltan, produced in countries including the Democratic Republic of Congo in socially and environmentally destructive ways. Every year, Britain ships some 500,000 tons of plastic to China to be recycled – although more of this material will soon have to be recycled directly in the UK, following an import ban imposed by China in early 2018.

In many prosperous middle-income countries, such as Brazil, India and China, rates of consumption for many millions of people in the middle and upper classes are as conspicuously high as they are in the West. In Jakarta, despite grinding traffic and an improving mass-transport system, rates of car ownership are still projected to increase dramatically. In Brazil, there is a vogue for mahogany and other timber decking among urban families which is leading to even greater pressure being placed on the country's forests.

Fortunately, it is possible to reduce levels of unnecessary consumption dramatically, at least for many billions of the world's people, while still enabling people to live happy and healthy lives. Indeed, evidence suggests life can be more pleasurable in the absence of such a strong focus on material possessions. One woman in Mill Valley, California, Bea Johnson, and her family of four, are following the five Rs: 'refuse, reduce, reuse, recycle and rot'. Her family stores all its food in jam jars, replenished from local wholesale shops rather than supermarkets. They compost all their food waste rather than dumping it at landfill. The entire wardrobe of each family member can fit into one small suitcase. Around the world, similar movements are gaining strength by the day, with a worldwide renaissance in thinking about and reducing people's and communities' footprint through bartering, trading, 'freecycling' (offering up materials online for people in the community to collect free of charge) and reusing materials. Together, then, we can decide knowingly to have fewer children, to eat less meat and to consume less stuff: this will all help a great deal. We can live lightly and respectfully on the face of the earth, and we must. There is so much each of us can do.

Action

My first nine paths set out what we need to do to put the planet on a better track. Renewable energy powers our economy. The world's forests are protected and restored. We are managing our soils for the long term. We are protecting vital fresh water and cherishing our biodiversity. We are also saving and restoring the ocean, as well as living in more sustainable cities. We are reducing waste of all kinds and establishing a circular economy. Finally, we are advancing ambitious and clear-sighted policies to enhance reproductive health care and rights, reduce excessive consumption and establish a healthier diet.

My final path is to bring about a world in which action on all of these nine areas is dramatically scaled up, fast. Just as after the Second World War, when the Marshall Plan enabled the reconstruction of a ravaged European continent, so too does the world now need an equivalent global effort to address our environmental predicament – a Marshall Plan for our times.

Path Ten

To restore the earth, and put the planet on a better course, we will need new and decisive shifts in the areas of economics, politics, the law, business, faith, the media and social networks, technology and innovation, and individual and societal ethics. We will also need a new story, or set of stories, by which to live. In my tenth and final path, then, I set out a vision for a global environmental Marshall Plan based on transformative shifts in these nine areas.

I. ECONOMICS

Our shared understanding of what makes for a good, fair and sustainable economic system is going to have to change if we are to put the world on a new track. Most contemporary economic discussions still give scant attention to climate change and other environmental concerns. And yet these same economics are at the heart of every country's decision-making and view of the world. The world's economists need to become the strongest advocates for the paths set out here, starting with the finance ministers of today, while also ensuring that the next generation of young economists are much more thoughtful about the environment than today's.

A former colleague at Oxfam, Kate Raworth, has set out a powerful vision of a new kind of economic system based on a fundamentally more equitable distribution of – and access to – the world's natural resources. Given our global obsession with economic growth, we need to gain a much better understanding of the underlying natural wealth which enables this growth. Kate's reformed economic system would give due weight to the global economic benefits of protecting the Amazon rainforest, for example, or removing subsidies supporting the use of fossil fuel. Her book *Doughnut Economics* should be mandatory reading for every finance minister and economics student the world over.

A critical part of winning the economic argument in favour of this book's recommendations is to demonstrate that addressing climate change and protecting the natural world is beneficial to economic growth, as well as to people's quality of life. There is bountiful evidence that this is the case, whether in the number of jobs generated by renewable energy or forest landscape restoration, or in the strong relationship between enhanced climate action and economic growth recorded by many countries. Despite such countries securing economic growth while reducing their carbon emissions, there is still much to be done to convince political leaders, whose default is to focus on the short-term. In particular, we need finance ministers to become visionary advocates for the paths set out here. A few of them already are, as is Christine Lagarde, the head of the IMF, but the majority of the world's financial elite tend to consider climate and the environment as an afterthought, if at all.

A proper price on carbon, air pollution and other environmentally damaging outputs, which reflects their true economic impact in terms of health or clean-up costs, would make a tremendous difference. $100 per ton of carbon emitted is widely accepted as an appropriate starting price, to be ratcheted up over time (although some argue that this should be much higher, and it will vary by country and region). Encouragingly, as described earlier, over 65 countries have already implemented some form of carbon pricing regime, China included. This trend will only increase, but it must do so quickly and at scale – and the price alighted on must be sufficient to lead to real change.

Dissatisfaction with the failures of the current economic system, due to rising inequality and people being left behind, has led to greater nationalism, political division and the rise of populism. In the US, unemployment, the financial crisis, economic marginalisation and inequality played a critical part in the election of Donald Trump, as similar underlying forces had led to Brexit months previously. In both cases, politicians were able to tap into these grievances to create a powerful political upset. The political crisis which has arisen in part due to the failings of the current system is an opportunity for a new and better economic model: one which not only leads to greater equality, but which restores and rejuvenates the earth. Around the world, countries, companies and societies are exploring such economies – and as we have seen, there are many encouraging examples of where a better economic system has been shown to be possible. But such examples need to become the rule, rather than the exception. Economics is at the heart of the better world we seek.

2. POLITICS

But societies are not moved by economics alone: we will also need a new politics capable of meeting the climate and sustainable development goals of our time. I dream of a much better, kinder and more generous politics, capable of uniting people, spanning the political divide and overcoming division. Part of the challenge will be to rise above the bipartisan divide which has recently bedevilled climate and environment action in the US, for example, but which also threatens many other countries, including the UK and Brazil. We should be reminded that no political orientation has a monopoly on positive environmental action: parties of both the left and right have been responsible for many environmental achievements, as well as a lot of destruction. A number of Republican presidents, including Nixon and Roosevelt, were responsible for some of America's most striking environmental laws and legacies, such as the National Parks system or the Environmental Protection Agency – now beleaguered in the Donald Trump era. Mao's China, and indeed Communist Russia, was as destructive of the environment as the fossil fuel lobby of the American libertarian right has been.

The UK's 2008 Climate Change Act, which sets a long-term, national low-carbon goal, was strongly supported by a majority of both Labour and Conservative Members of Parliament, as well as by all the other political parties. British environmental NGOs, which have at least 4 million members (more than all the political parties combined), worked carefully to build a national consensus on the environment which largely, mercifully, still holds – despite the broader divisions in our society, which came to a head during the EU referendum in 2016.

Environmentalists need to engage politically in astute and strategic ways with the political class and the public at large, using the kinds of arguments which hold force today: jobs, economics, growth, national pride. In the US Presidential election in 2016, Hillary Clinton never successfully confronted Donald Trump's rhetoric about job losses for coal miners with the facts about job creation in solar and renewable energy: there are now 260,000 solar workers in the US, five times the number of coal miners.

President Emmanuel Macron is arguably one positive example of what we need from political leaders on the global stage, exalting the virtues of the Paris Agreement and positioning France as a climate leader, tackling President Trump in private and in public on the issue in no uncertain terms. Chancellor Merkel, herself a scientist, has been instrumental in Germany's energy policy, the *Energiewende* – the subject of much critique from German environmentalists. President Xi Jinping has done the same for China, leading on China's much-heralded (and hoped for) vision of an 'Ecological Civilisation'. Prime Minister Modi in India has made ambitious commitments on solar energy and on cleaning up the River Ganges, although his environmental record is problematic in other ways. President Paul Kagame in Rwanda has been active in seeing through the country's recent ban on plastic bags as well as an enlightened commitment to better soil and forest management. Prime Minister Trudeau in Canada has been positive about climate and marine conservation action, although equivocal on phasing out tar sands – a highly polluting form of fossil fuel – and oil pipelines. President Vladimir Putin of Russia, while apparently a passionate conservationist when it comes to tigers, has been highly critical about the need to act on climate change. We will need ambitious political environmental leadership across the world in the years ahead: and for this to happen, in democracies at least, we will need people to clamour for it, and to vote for environmentally minded leaders.

3. THE LAW

The law is the single most powerful means by which to call companies and countries to account, on a whole range of issues from air pollution and deforestation to pesticide use, as well as to drive long-lasting environmental change and to ensure that commitments, such as the Paris Agreement, are honoured.

One of the most inspiring interviews I carried out for this book was with James Thornton, Founder and Director of Client Earth. He was voted one of the UK's top ten environmental heroes in 2017, and in 2012 by the *New Statesman* as one of the ten people who could change the world. A practising Buddhist, James is a compelling man: softly spoken, articulate, charming and focused. When we talked, he ran through impressive cases in which the organisation has brought about major legislative victories on air pollution, illegal timber, marine fisheries and climate action across Europe, the US and beyond. There seems to be no end to the potential power of the law to protect the environment, providing that organisations are equipped with the resources and budgets they need.

In 2017, Client Earth led at least 12 campaigns which hit the headlines. In January, they demonstrated that the UK was in breach of its emergency pollution ceilings, which led to significant public outrage and the promise of a legislative response from the government. In February, they tackled illegal levels of air pollution in Italy, leading the authorities to commit to a new air quality plan. In April, they played a significant role in the new EU law leading to tougher pollution standards for coal plants. In May, they published a new legal guide for forest communities in Liberia. The organisation was registered in China in June, a rare feat for a Western NGO, and has been training judges on environmental law there since. In September, it tackled the UK on air pollution in children's playgrounds. In October, it made Belgium the first country to be brought to court for failure to use the EU Timber Regulation, which prevents the selling of illegally harvested timber. Throughout the year, Client Earth took action against coal plants in Italy, Greece, Bulgaria and Spain. In November, in partnership with six other organisations, it won legal action against Poland's planned three-fold increase of logging in the

Białowieża Forest, an iconic old growth forest of oustanding beauty and biodiversity. And finally, in December 2017, Client Earth convinced the UK's Department of Work and Pensions to change the law to establish pension schemes' obligation to consider environmental risks such as climate change.

When I first heard James speak about the work of Client Earth, at a meeting in Oxford, it felt like an epiphany. In his quiet, committed, compelling way, James argued that there are no limits to what the law can do for the environment – and he is right.

4. BUSINESS

The private sector has such power and holds such sway that it could also lead the way in bringing about a better world, even in the absence of governments. It can also be deeply obstructive, as has been the case historically with oil companies funding climate change denial for decades. There is an urgent need to put sustainability challenges at the heart of businesses' growth plans. The businesses of tomorrow need to be circular, regenerative, equitable, visionary, fleet of foot, and inspired by a deep commitment to social and environmental action. In other words, they need a 'licence to operate'. Their own employees will expect it.

Unilever, under Paul Polman, has shown over the past ten years that a company can become substantially more sustainable if it really makes a concerted effort to do so (and, in Unilever's case, if it builds on the decades of thinking and ethical concern which preceded Paul's arrival). Others are also making valuable efforts: BT is working to cut greenhouse gas emissions from its supply chain by 29 per cent by 2030. Mars, Inc. established a 'Sustainable in a Generation Plan' to freeze its ecological footprint; it is also giving serious thought to the issue of ensuring fair pay for rural farmers in its supply chains. Marks and Spencer has a sustainability plan which has been a feature of the company for a decade or more, led by the irrepressibly enthusiastic and clear-sighted Mike Barry. Anand Mahindra, the Indian tycoon and chair of the Mahindra Group, said at Davos in 2018 that 'climate change is in fact the next century's biggest financial and business opportunity'. More than 330 companies have committed their operations to meet science-based climate and sustainable development targets. We need many thousands more companies, from global multinationals to small start-ups, to replicate these aims.

The new business models the world needs generate value for a company through social inclusion and environmental sustainability, and focus on a 'triple bottom line' – people, planet, profit – rather than solely on profit. In recent years, it has been encouraging to watch how businesses such as Unilever speak about the importance of using their business to achieve real social and environmental good. Paul Polman has been one of the most consistent, outspoken and visionary business advocates for sustainable development, and was an instrumental figure in achieving the climate and Global Goals of 2015.

I interviewed Paul a few months prior to Oli's birth, on the top floor of the Unilever London office overlooking St Paul's Cathedral and the Tate Modern. He liked to think of himself as a 'prisoner of hope', he said, citing Desmond Tutu's phrase, and stating 'that negative thoughts and cynicism don't get you anywhere'. He believes very deeply that 'it will be possible to irreversibly eradicate poverty in a sustainable and equitable way, while at the same time living within planetary boundaries.' And he argues that 'Paris sent a clear signal, and that there has been a rapid acceleration of action since then, including as a result of the fossil fuel divestment movement which has demonstrated that companies have $35 trillion worth of assets on their balance sheets which cannot be used if we are to meet the Paris Agreement, and which are therefore stranded.'

The recent Business & Sustainable Development Commission, which Paul co-chaired, argued that the actions required to meet the Global Goals would generate some $12 trillion, and that any savvy business should position their company to benefit from such investments. He also continues to believe that businesses like Unilever can be an active force for good. They can eliminate deforestation from palm oil, soya, cocoa, beef and timber in their supply chains, and indeed have made a commitment to do so by 2020. They can reduce food and materials waste in their factories and farms. They can promote the circular economy and greater recycling. They can speak up against negative advertising. Unilever's 'bouillon', fortified with iron, can reduce child mortality, helping to avoid 3 million children dying of malnutrition and diarrhoea every year; their Lifebuoy antibacterial soap can enhance health and sanitation and help get rid of infectious diseases.

In summary, every aspect of a company's huge and varied operations should contribute to sustainability and human development. In the months that followed our interview, Unilever was subject to a hostile takeover bid, and a key part of the company's pushback was based on the broader values they seek to embody. The world and the company's shareholders appeared to rally in a similar vein.

Paul was proud that precisely because of its sustainability commitment, Unilever is one of the most popular companies on university career websites and for young graduates. 'At a time when the average age of publicly traded companies has gone down from 65 to 17, businesses only really flourish if they have a value for society and a sense of purpose,' he said. Paul has shown that there is another way of doing business, and that setting the bar high can pay handsome dividends. But there are many other examples too, and a whole new cadre of young business leaders is doing things differently, with a much greater commitment to sustainability. Businesses can play a huge part in transforming our world, and indeed they will have to.

5. FAITH

The immense power and sheer numbers of adherents to the world's religions mean they are already at the heart of hundreds of positive actions to address environmental challenges. Most if not all of the world's major religions have respect for nature and environmental stewardship at their core, based either on their creation stories or through their focus on simplicity, moderation and the Buddhist conception of 'right livelihoods'.

This is not to say that some strands of faith and faith leaders are not at odds with environmental concerns. In the US, there are parts of the evangelical Christian community which believe climate change is a hoax and a left-wing conspiracy; that evolution did not happen; and that apocalypse of any kind, including climatic, will effectively herald the Second Coming of Christ and the redemption of the world. In such communities, there is clearly considerable work to be done to bring leaders and faith groups on side with the environmental cause.

In 2015, Pope Francis instigated a remarkable effort on behalf of the Catholic church to tackle climate change and environmental degradation, in part through a Papal Encyclical, *Laudato Si'*, or *On Care for our Common Home*. I read *Laudato Si'* in a single three-hour sitting one afternoon on the eve of its publication, and was astonished by the breadth and acuity of its vision. Almost all of the issues in this book (with the exception of reproductive health care and rights, which are alas barely mentioned – and when they are, in a fairly dismissive fashion) are pithily summarised in the Encyclical. It has been widely debated and analysed across Catholic congregations, although its findings and vision are by no means universally accepted. Analysts of the Paris Agreement believe the Encyclical played its part in securing the conditions for a strong outcome from the talks. We need very much more action of this kind from the world's religious leaders.

Not long before Oli's birth, I was fortunate enough to interview one of the people I most admire: poet, writer and former Archbishop of Canterbury, Rowan Williams, who is now Master of Magdalene College, Cambridge, and Chair of the UK-based charity Christian Aid. At the offices of Christian Aid, I asked him how we might construct a narrative which would unite rather than divide us, overcoming the discord and division of these times. Rowan argued that much of the answer was to do with 'recognising other human beings and being recognised by them: putting someone at the centre of our thoughts'. We also need to accept 'we will only be secure if our neighbours are secure'. Rowan described how the Hebrew Scriptures saw 'the world we inherited as a garden, in which the earth works at its own rate with seasons, succession', such that we ourselves needed 'to learn the ways of the world, its rhythm and fabric. This is something Christians, Jews and Muslims all share.'

He also referenced the concept in Hebrew scripture, 'that our environment is lent to us, we don't possess it': that there is 'a need to let things go; that there is a cyclic character to life; we have to step back to let nature be itself; to re-establish its own balance'. He wondered whether a myth for our time would be about 'gift': 'we live in an environment of actual and potential gift. What is the gift waiting to be given? It comes in unexpected ways, underpinned by the basic conviction crystallised in the self-communication of the divine. This is not a sentimental observation from Wordsworth. We are so constituted that we are ourselves attuned to this intelligence; we grow as human beings to be receptive; we're not strangers.' If we were more fully aware, as a result of our religious traditions or not, of the remarkable gift inherent in being alive on a planet of this beauty and abundance, we would surely behave more respectfully and with greater care towards it.

I asked Rowan what God might make of our current ecological predicament. He answered that he might say: 'Haven't you been listening?' Rowan talked about the importance of paying attention to the rhythms of the natural world, just 'as one pays attention to one's own body'; and that the ancient traditions focused on the material world, on ritual and sacrament, arguing for 'a reverent use of the things around us: let us be mindful and thoughtful'. He quoted approvingly a French saint in the eighteenth century who said to an aristocratic lady who hoped to enter into the sainthood that she might do so, provided that she learned to 'cook more slowly; talk more slowly; and eat more slowly'.

On the role of individuals, Rowan argues:

Path Ten

It is very important that there be things people can do. We're not all doomed. Low energy living is one. Taking time is another. I was very happy to hear that there was a Slow Food Movement. A heightened level of simple thoughtfulness is important. Then there is advocacy, whether on the ethics of a pension fund; asking difficult questions of others' economic projects in terms of their environmental integrity; to share an individual's experience: this is the least we can do. Doing things for their own sake is important. In some moods, to those who say we are all doomed, I am minded to say 'so what': there are good and bad ways to live; let not there be a counsel of despair; let us live well; this is the ethical thing to do. I hope it's not too late to turn things around, but even if it is, I do this finally because it is a good way of being in the world.

Finally, I asked Rowan whether he is optimistic. 'I'm not optimistic, I'm hopeful,' he said. 'The cards are heavily stacked; especially with the current politics, the retreat from thinking through our options in the post-truth climate by people who should know better.' He is however heartened by the 'level of awareness and ownership of the environmental agenda in the under-35s'. Corporate practice is more under scrutiny than before, and 'the much-maligned social media' provides a good opportunity to spread stories of good practice. In summary, 'We have the tool cabinet and the tools we need; if we pretend otherwise, we are not honouring future generations. We need to honour what's been given to us and honour the generations to come; to act with corporate and political intelligence, and the cooperation that comes with it.'

Rowan's lucid environmental vision is matched by similar thinking from other spiritual leaders, from the Dalai Lama to Rabbi Jonathan Sacks, and other faith traditions, including Islam, Sikhism, Daoism and Hinduism. But there is much more still to be done to ensure that the world's faith communities mobilise their followers. The Alliance of Religions and Conservation is one group dedicated to greater environmental consciousness and action in the faith community, whether through supporting the EcoSikh group in its successful campaign to ensure that the Golden Temple in Amritsar serves organic food to its 100,000 daily visitors; or Shinto shrines in Japan to manage their forests more sustainably and source only sustainably harvested wood for their 80,000 shrines. There are also important efforts under way to encourage the faiths to divest from fossil fuels and invest proactively in renewable energy, with encouraging new announcements frequently made by the Church of England, the Vatican and others. If the world's faiths truly came on board with the agenda described in this book we'd be a lot closer to the (proverbial) promised land.

6. THE MEDIA AND SOCIAL NETWORKS

We live, as Sir David Attenborough said, in a time of unprecedented global awareness about humanity's impact on the natural world, and this has been achieved as a result of media coverage, documentaries, BBC series and the like. The impact of Attenborough's *Blue Planet II* has been extraordinary, both in terms of generating public awareness and in lending weight to campaigns such as one to extend the UK's 'Blue Belt' of Marine Protected Areas or to encourage the UK to adopt a bottle-and-can deposit return scheme. *Blue Planet II* has prompted 90 per cent of viewers to change their shopping and consumption habits. But it remains the case that at a time of 'fake news', more and more ideologically divided and polarised news coverage, and the absence of press freedom in many illiberal states, not enough people have the information they need to make good environmental decisions and push for change. Social networks, however, present new opportunities for environmental action, whether through dissemination of information or effective campaigns. In 2015, a video by the Chinese activist Chai Jing, *Under the Dome*, about air pollution in Beijing was viewed over 150 million times within three days, and had been seen a further 150 million times by the end of the first week – at which time it was taken offline. Companies respond fast to well-targeted social media campaigns, such as the global petition and campaign website, Avaaz, and the like.

The challenge then becomes to turn a more informed and concerned population into a more politically effective one, as well as to ensure that increased levels of concern lead to greater individual action. If not managed well, increased environmental information can lead to inertia and indifference: people become fatalistic and wonder if there is any point. It is therefore important to ensure that environmental programmes and articles focus on the solutions, both at the local level and in terms of what is required globally, rather than merely describing the problem.

7. TECHNOLOGY AND INNOVATION

Many believe that technology is our strongest hope of a better future. We have seen that renewable energy is disrupting the hegemony of the oil and gas incumbents and providing energy to millions of people off the grid. Drones can be used to reforest distant landscapes and to monitor wildlife populations. In cities, innovations in public transport systems are providing alternatives to the car to hundreds of millions of people. And the circular economy and new approaches to manufacturing are occurring at an unprecedented rate.

There is in principle no limit to what might be achieved by new technologies, whether in cultured meat, improvements in desalination membranes, thermal depolymerisation, vertical farming, improved waste processing or in some of the more benign mooted forms of geoengineering, such as stimulating algal growth to absorb CO_2 emissions. The changes are happening astonishingly fast. In the case of energy, indeed, such is the pace of change that one might now refer to 'negatwatts' of electricity, instead of 'megawatts': electricity that, by means of energy efficiency, was not needed and therefore not generated in the first place. Sir Jonathon Porritt's *The World We Made* is a convincing illustration of how the world might look in 2050, if these technologies flourish in the ways they are likely to.

Technology will however not be our salvation on its own, and will bring negative aspects with it as well as positive ones, such as for people's jobs and livelihoods, as machines do more work than ever before. These developments need to be handled with care, and dystopian outcomes avoided wherever possible. And, even in the most optimistic technological scenarios, there will still be a need for substantial shifts in consumption and our management of the natural world. No single technology can protect elephants in the wild, prevent fresh water aquifers from being overexploited, or the rainforests from being cut down, in and of itself. But we do have the chance to harness the technological revolution to positive environmental ends, and so seize it we must.

8. EMPATHY AND EDUCATION

Ultimately, people need to love and value nature in order to protect it. Better education and more access to nature for children and communities will enhance the world's levels of individual and societal commitment to the environment and an ethic of care. To achieve such levels of consciousness takes time, political leadership, storytelling and art; and societies prioritising the environment at crucial historical moments. We need, urgently, to develop a sense of empathy and care for the natural world akin in strength and devotion to the love one feels for one's child.

From the UK to the Democratic Republic of Congo, from Brazil to Indonesia and from the US to Australia, there is a flourishing green movement comprising many millions of members. This movement has won, and continues to win, important battles. But we need societies as a whole to be as invested in these issues as these movements are: to move from the environment being a concern of only a relative few to becoming a universally shared one. Norms need to change: plastic straws need to be considered an unacceptable contributor to ocean plastic, rather than an indispensable part of a drink. As societies, the ambit of our ethical concern also needs to extend to a fuller appreciation of the wrongs of industrial animal farming, on both animal welfare and environmental grounds. Mahatma Gandhi had it right when he said: 'The greatness of a nation and its moral progress can be judged by the way its animals are treated.' We will need to expand our moral compass and begin to act as true custodians of the earth.

9. A NEW NARRATIVE

Finally, we need to find a way of telling a compelling,

new story about the sustainable planet we wish to bring into being. (This book, in its own modest way, is my attempt to do this.) Societies still largely operate on the basis of myths and stories, as Alex Evans describes in *The Myth Gap*. We either come up with our own set of compelling stories which win the public imagination, or others will: witness how Donald Trump got elected with a story all of his own, brutally effective, however problematic and fundamentally flawed.

We will need to capture people's imaginations with a hopeful and inspiring vision of an alternative, better future if we are to bring about the big global effort and Marshall Plan that I describe. I interviewed Alex, who has been a policy expert at the heart of many of the key international policy debates and processes of the past fifteen years. His argument is that we need to tell a story about the world which resonates with people, rather than coming across as something abstract and technocratic. This would be a vision based less on facts, dry arguments and statistics, and much more of an emotional narrative – or 'myth' – that people can be moved and enthralled by. The challenge is also to find a way of breaking out of our social media echo chambers to find a way, in Jo Cox's words, of 'uniting rather than dividing us'. At a time of such polarisation on so many issues, this is not easy to achieve, and yet more vital than ever before.

Myths, whether from the Bible, or *The Lord of the Rings*, or *Frozen*, one of Alex's daughter's favourite films, captivate us in ways drier arguments never can. The political parallels appear self-evident, in a 'post-truth' world in which Donald Trump regularly invents facts and stories. The challenge is to articulate a compelling story of the world we wish to see, to communicate that story widely and then to take immediate and daring steps to bring that world into being.

With concerted action in these nine areas, and a Global Marshall Plan to put the planet on a better course, it is my profound belief that a peaceful future for humanity and the natural world remains within our grasp. And so we must act, with the best of the human spirit, intelligence, morality and courage, to do everything we can to bring this world into being.

Hope

Oli arrived in the world in April 2017, in conditions of some distress: his lungs pumping faster than average, evidence that he was fighting off an infection acquired in the later stages of labour. Shortly after being born, the nurses took him to the neonatal ward of the Royal Free, where he spent his first two nights being looked after by an attentive paediatric nurse, housed in a little glass cubicle in which the temperature and oxygen levels were regulated, and his heartbeat and vital signs of life closely monitored.

I remember those stark few hours that night at Oli's side, willing him on, and not really knowing whether he would be all right. The nurse seemed calm and confident, and all the knowledge and technology of modern medicine were visible in the room, and deeply reassuring. In the days that followed, as Oli began to recover and fight off his infection with the help of antibiotics, Davina and I were constantly moved by the wonderful nurses and doctors who cared for him. We will always remember the help they gave to his exhausted parents, and the sheer quality of care and attention that he received. Above all, we felt, there was a remarkable ethic of life and care in evidence everywhere in the hospital. Five days after Oli was born, Davina and I returned home from the hospital with Oli in our arms – and our lives together began.

I describe all of this because it seems to me now (as it did in the hospital at the time) that the parallels between what we need to do to restore the earth and put the planet on a better course, and how the hospital cared for Oli, are strikingly clear. The incubator in which Oli spent his first and second nights, with its climate, oxygen levels and temperature regulated for his survival, is emblematic of the planetary environment that we must maintain for humanity's and the earth's well-being. But above all it was the extraordinary duty of care from the people caring for our son that moved me. Surely what the world needs more than anything else, I felt, is a similar duty of care to the planet.

Can we not replicate the Hippocratic oath, for the relationship between doctor and patient, in humanity's relationship with the natural world? My deep love for my son translates into a deep and constant concern for the world into which I've brought him. Ultimately, surely, the solution is to apply the same ethic of care that we saw at the Royal Free to the planet.

Michael McCarthy has argued that 'ordinary people's feelings contain the beginnings of political will'. Despite the scale of the human enterprise, it must be possible to protect the natural world which remains and restore much of what we have lost. I hope this book plays some part in encouraging people to think politically and to contribute to a restored earth.

Sir Jonathon Porritt, who stated when we met that he was 'a fully paid up subscriber to the genius of the human species', contended that our potential to do good was extraordinary; and that our capacity to harness 'a massive force of creativity, tenderness, compassion, tolerance, pluralism, empathy; rather than xenophobia, hatred, and exclusivity' was undiminished. 'If we provide the right conditions for young people, everything is possible,' Jonathon believes.

As Oli, and all who are born alongside him in this world of ours, begins to make his way, all I can hope is that we will – all of us – wake up to the scale of the challenges before us, and act as never before.

There are many encouraging signs that we are on the cusp of the most astonishing transition towards a more sustainable and harmonious way of living within the natural environment. But we could, equally, lose sight of what is important and continue to damage the natural world and the biosphere in the ways we do today. The future rests in our hands, and we must do all that we can. Given half a chance, the planet and the natural world have a remarkable capacity to recover. I believe very deeply that we can put things right. We know so much about what needs to be done. If we could go to scale on the approaches I have described in this book, we would make the most remarkable difference. And so we must.

Further Reading

MAIN SOURCES

Berry, Wendell, *The World-Ending Fire: The Essential Wendell Berry*, selected and introduced by Paul Kingsnorth, Penguin, UK, 2017

Crane, Nicholas, *You Are Here: A brief guide to the world*, Weidenfeld & Nicolson, UK, 2018

Evans, Alex, *The Myth Gap: What Happens When Evidence and Arguments Aren't Enough*, Eden Project, UK, 2017

Faciolince, Héctor Abad, 'A Rationalist in the Jungle', *Granta Magazine*, UK, 2003

Pope Francis, *Laudato Si': On Care for Our Common Home*, Catholic Truth Society, UK, 2015

Gore, Al, *The Future*, WH Allen, UK, 2014

McCarthy, Michael, *The Moth Snowstorm: Nature and Joy*, John Murray, UK, 2016

Pollan, Michael, *In Defence of Food: The Myth of*

Nutrition and the Pleasures of Eating, An Eater's Manifesto, Penguin, UK, 2009

Porritt, Jonathon, *The World We Made: Alex McKay's Story from 2050*, Phaidon, UK, 2013

Roy, Arundhati, *The Cost of Living: The Greater Common Good and the End of Imagination*, Flamingo, UK, 1999

Thesiger, Wilfred, *The Marsh Arabs*, Penguin, UK, 2007

SELECTED FURTHER READING

GENERAL

Diamond, Jared, *Collapse: How Societies Choose to Fail or Survive*, Penguin, UK, 2011

Flannery, Tim, *Atmosphere of Hope: Solutions to the Climate Crisis*, Penguin, UK, 2015

Harari, Yuval Noah, *21 Lessons for the 21st Century*, Jonathan Cape, UK, 2018

—, *Homo Deus: A Brief History of Tomorrow*, Vintage, UK, 2017

—, *Sapiens: A Brief History of Humankind*, Vintage, UK, 2015

Hawken, Paul, *Drawdown: The Most Comprehensive Plan Ever Proposed to Reverse Global Warming*, Penguin, UK, 2018

Juniper, Tony, *What Has Nature Ever Done for Us? How Money Really Does Grow on Trees*, Profile, UK, 2013

Leggett, Jeremy, *The Winning of the Carbon War: Power*

and Politics on the Front Lines of Climate and Clean Energy, Crux, UK, 2018

Lewis, Simon, and Maslin, Mark, *The Human Planet: How We Created the Anthropocene*, Pelican, UK, 2018

Lovelock, James, *A Rough Ride to the Future*, Penguin, UK, 2015

Lynas, Mark, *The God Species: How Humans Really Can Save the Planet*, Fourth Estate, UK, 2012

Monbiot, George, *Feral: Rewilding the Land, Sea and Human Life*, Penguin, UK, 2014

Morton, Oliver, *The Planet Remade: How Geoengineering Could Change the World*, Granta, UK, 2016

Pearce, Fred, *Fallout: A Journey Through the Nuclear Age, From the Atom Bomb to Radioactive Waste*, Portobello, UK, 2018

Rosling, Hans, *Factfulness: Ten Reasons We're Wrong About The World – And Why Things Are Better Than You Think*, Sceptre, UK, 2018

Sachs, Jeffrey, *The Age of Sustainable Development*, Columbia University Press, US, 2015

Schumacher, Eric, *Small Is Beautiful: A Study of Economics as if People Mattered*, Vintage, UK, 1993

Weisman, Alan, *The World Without Us*, Virgin, UK, 2008

PATH ONE: RENEWABLES

Goodall, Chris, *The Switch: How Solar, Storage and New Tech Means Cheap Power for All*, Profile, UK, 2016

Helm, Dieter, *Burnout: The Endgame for Fossil Fuels*, Yale University Press, US, 2017

MacKay, David, *Sustainable Energy – Without the Hot Air*, Green, UK, 2009

Sivaram, Varun, *Taming the Sun: Innovations to Harness Solar Energy and Power the Planet*, MIT Press, US, 2018

PATH TWO: FORESTS

Davis, Wade, *One River: Explorations and Discoveries in the Amazon Rain Forest*, Vintage, UK, 2014

Juniper, Tony, *Rainforests: Dispatches from Earth's Most Vital Frontlines*, Profile, UK, 2018

Seymour, Frances, and Busch, Jonah, *Why Forests, Why Now?: The Science, Economics, and Politics of Tropical Forests and Climate Change*, Center for Global Development, US, 2014

PATH THREE: SOIL

Berry, Wendell, *The World-Ending Fire: The Essential Wendell Berry*, selected and introduced by Paul Kingsnorth, Penguin, UK, 2017

Carson, Rachel, *Silent Spring* (1962), Penguin, UK, 2000

Conway, Gordon, *One Billion Hungry: Can We Feed the World?*, Comstock Publishing Associates, UK, 2012

Kumar, Satish, *Earth Pilgrim*, Green Books, UK, 2009

—, *Soil, Soul, Society: A New Trinity for Our Time*, Leaping Hare Press, UK, 2015

Lynas, Mark, *Seeds of Science: Why We Got It So Wrong on GMOs*, Bloomsbury, UK, 2018

McMahon, Paul, *Feeding Frenzy: The New Politics of Food*, Profile, UK, 2013

Montgomery, David, *Dirt: The Erosion of Civilizations*, University of California Press, US, 2012

Montgomery, David, and Biklé, Anne, *The Hidden Half of Nature: The Microbial Roots of Life and Health*, W. W. Norton, US, 2015

Tree, Isabella, *Wilding: The Return of Nature to a British Farm*, Picador, UK, 2018

Tudge, Colin, *Six Steps Back to the Land: Why We Need Small Mixed Farms and Millions More Farmers*, Green Books, UK, 2016

PATH FOUR: WATER

Mallet, Victor, *River of Life, River of Death: The Ganges and India's Future*, OUP, UK, 2017

Pearce, Fred, *When the Rivers Run Dry: What Happens When Our Water Runs Out?*, revised and updated edition, Beacon, US, 2018

Schwartz, Judith, *Water in Plain Sight: Hope for a Thirsty World*, St Martin's Press, US, 2016

PART FIVE: BIODIVERSITY

Cocker, Mark, *Our Place: Can We Save Britain's Wildlife Before It Is Too Late?*, Jonathan Cape, UK, 2018

Cocker, Mark, and Tipling, David, *Birds and People*, Jonathan Cape, UK, 2013

Further Reading

Kolbert, Elizabeth, *The Sixth Extinction: An Unnatural History*, Bloomsbury, UK, 2015

McCarthy, Michael, *The Moth Snowstorm: Nature and Joy*, John Murray, UK, 2016

Pearce, Fred, *The New Wild: Why Invasive Species Will Be Nature's Salvation*, Icon, UK, 2016

Pollan, Michael, *The Omnivore's Dilemma: The Search for a Perfect Meal in a Fast-Food World*, UK, Bloomsbury, 2011

—, *The Botany of Desire: A Plant's-eye View of the World*, Bloomsbury, UK, 2002

Shanahan, Mike, *Ladders to Heaven: The Secret History of Fig Trees*, Unbound, 2016

Sheldrick, Dame Daphne, *Daphne: An African Love Story, Love, Life and Elephants*, Penguin, UK, 2013

Solnit, Rebecca, *Savage Dreams: A Journey into the Hidden Wars of the American West*, University of California Press, US, 2014

Willis, Kathy, and Fry, Carolyn, *Plants: From Roots to Riches*, John Murray, UK, 2014

Wilson, Edward O., *Half-Earth: Our Planet's Fight for Life*, Liveright, UK, 2017

—, *The Future of Life*, Abacus, UK, 2003

—, *The Diversity of Life*, Penguin, UK, 2001

—, *Consilience: The Unity of Knowledge*, Abacus, UK, 1999

Wulf, Andrea, *The Invention of Nature: The Adventures of Alexander von Humboldt, The Lost Hero of Science*, John Murray, UK, 2016

Further Reading

PATH SIX: OCEAN

Hoare, Philip, *Leviathan*, Fourth Estate, UK, 2009

Honeyborne, James, and Brownlow, Mark, *Blue Planet II*, BBC Books, UK, 2017

Kurlansky, Mark, *Cod: A Biography of the Fish That Changed the World*, Vintage, UK, 1999

McCallum, Will, *How to Give Up Plastic: A Guide to Changing the World, One Plastic Bottle at a Time*, Penguin, UK, 2018

Roberts, Callum, *The Ocean of Life: The Fate of Man and the Sea*, Penguin, UK, 2013

Siegle, Lucy, *Turning the Tide on Plastic: How Humanity (And You) Can Make Our Globe Clean Again*, Trapeze, UK, 2018

PATH SEVEN: CITIES

Barber, Benjamin, *Cool Cities: Urban Sovereignty and the Fix for Global Warming*, Yale University Press, US, 2017

Bloomberg, Michael, and Pope, Carl, *Climate of Hope*, St Martin's Press, US, 2017

Glaeser, Edward, *Triumph of the City*, Pan, US, 2012

Goodell, Jeff, *The Water Will Come: Rising Seas, Shrinking Cities, and the Remaking of the Civilized World*, Black, Australia, 2018

Montgomery, Charles, *Happy City: Transforming Our Lives Through Urban Design*, Penguin, UK, 2015

Further Reading

PATH EIGHT: WASTE

Lymbery, Philip, and Oakeshott, Isabel, *Farmageddon: The True Cost of Cheap Meat*, Bloomsbury, UK, 2015

Stuart, Tristram, *Waste: Uncovering the Global Food Scandal*, Penguin, UK, 2009

Stuchtey, Martin, Enkvist, Per-Anders, and Zumwinkel, Klaus, *A Good Disruption: Redefining Growth in the Twenty-First Century*, Bloomsbury, UK, 2016

PATH NINE: PEOPLE

Dyson, Tim, *Population and Development: The Demographic Transition*, Zed, UK, 2010

Jackson, Tim, *Prosperity Without Growth: Foundations for the Economy of Tomorrow*, 2nd edition, Routledge, UK, 2016

Johnson, Ben, *Zero Waste Home: The Ultimate Guide to Simplifying Your Life*, Penguin, UK, 2016

Sen, Amartya, *Development as Freedom*, OUP, UK, 2001

PATH TEN: ACTION

Elkington, John, and Zeitz, Jochen, *The Breakthrough Challenge: 10 Ways to Connect Today's Profits with Tomorrow's Bottom Line*, Jossey-Bass, UK, 2014

Ghosh, Amitav, *The Great Derangement: Climate Change and the Unthinkable*, University of Chicago Press, US, 2016

Further Reading

Lent, Jeremy, *The Patterning Instinct: A Cultural History of Humanity's Search for Meaning*, Prometheus, US, 2017

Marshall, George, *Don't Even Think About It: Why Our Brains Are Wired to Ignore Climate Change*, Bloomsbury, US, 2015

Raworth, Kate, *Doughnut Economics: Seven Ways to Think Like a 21st-Century Economist*, Random House, UK, 2018

Thornton, James, and Goodman, Martin, *Client Earth*, Scribe, UK, 2017

EPILOGUE: HOPE

Balmford, Andrew, *Wild Hope: On the Front Lines of Conservation Success*, University of Chicago Press, US, 2012

Monbiot, George, *Out of the Wreckage: A New Politics for an Age of Crisis*, Verso, UK, 2018

Townsend, Solitaire, *The Happy Hero: How to Change Your Life by Changing the World*, Unbound, UK, 2017

Organisations

There are so many organisations doing remarkable work on the environment. Here are ten whose work and vision inform this book:

Greenpeace
Rainforest Foundation Norway
Soil Association
WaterAid
WWF
Surfers Against Sewage
C40
Feedback
UNICEF
Client Earth

Acknowledgements

My first vote of thanks is to Davina for all her support, love, wisdom, patience, understanding and encouragement. I am so deeply grateful to her, more than I can say; I hope she and Oli enjoy the book.

My very grateful thanks are also due to my family for all their generous encouragement and support: Mum and Dad, who have been wonderful rocks throughout the years and adventures described in this book, and Kerstin, Nick and Kirsten, Anna and Nick, Ben and Victoria, Tom and Emmy, Susan and Christopher, Rob and Katrina, Barnaby and Nicole, Oliver and Magnolia, Luke and Marija, Adrian and Cassandra, Anthony and Sara, Judy, Fred and Lin, and Alex and Mark.

Many thanks to all my dear friends and former colleagues at the Prince of Wales' International Sustainability Unit, for their generosity of spirit, commitment, collegiality and kindness throughout the process of writing this book, as well as for their guidance and comments on the text at various points. All the ideas and issues discussed in the book are ones on which the team

Acknowledgements

has worked over the past ten years, and their thinking and insight informs every chapter – although I should say that all errors of fact, interpretation or emphasis are entirely my own. The same applies to my wonderful new employers, the World Resources Institute, whose daily work and immense expertise in all the areas of this book are also a daily inspiration.

By the same token, I would like to put on record my sincere gratitude to HRH The Prince of Wales for his long-standing leadership and inspiration on the environment, and for the privilege and opportunity that it was to work for the ISU in pursuit of decisive international action to address a set of causes about which he has been so passionate for such a long time.

I am very grateful to all those interviewed for the book for sharing their considerable knowledge, experience, ideas and wisdom: their thinking has had a huge impact on the ideas shared here. The full list includes: May Al-Karooni, Sir David Attenborough, Monique Barbut, Owen Barder, Mike Barry, David Bent, Paul Bodnar, Sir Gordon Conway, Professor Sebsebe Demissew, Professor Tim Dyson, John Elkington, Alex Evans, Christiana Figueres, Barbara Frost, Nora Fyles, Tara Garnett, Leo Hickman, Martin von Hildebrand, Isabel Hilton, Tony Juniper, Sir David King, Satish Kumar, Rachel Kyte, Jeremy Leggett, Tony Long, Michael McCarthy, Professor Georgina Mace, Lord Peter Melchett, Dr David Nabarro, David Nussbaum, Paul Polman, Sir Jonathon Porritt, Matt Rand, Daniel Raven-Ellison, Simon Reddy, Louise Rouse, John Sauven, Erik Solheim, Andrew Steer, Achim

Acknowledgements

Steiner, Lord Nicholas Stern, Tristram Stuart, Martin Stuchtey, James Thornton, Sir Crispin Tickell, Lord Adair Turner, Jakob von Uexküll, Professor Peter Wadhams, Mark Watts, Shelagh Whitley, Baron Williams of Oyster-mouth and Professor Kathy Willis. I am most grateful to them all, as well as to many other friends and colleagues not interviewed for the book but whose ideas nevertheless form the backdrop to all that is discussed here.

I would also like to thank Justin Mundy, Tony Juniper, Tom Phillips and Joe Whitlock-Blundell for their judicious comments on an early draft of the book proposal (and, in Justin's case, various drafts thereafter), and Tony for being a writing mentor and inspiration throughout (in the time taken to write this book, he dashed off another master-piece, his latest book on rainforests). Excellent, perceptive and stimulating comments on later drafts were also received from David Bent, two friends both by the name of Tom Harrison, Marjory-Anne Bromhead, Gill Shepherd, Kate Upshon, Tom Hay and my dear cousins Oliver and Barnaby Phillips. Markus Petz did a helpful proofread on the eve of submission of the manuscript.

I am truly grateful to my kind and solicitous editor Philip Connor for all his support and intelligent, empa-thetic encouragement throughout the project; the book emerged from a much longer manuscript thanks to his skilful work. Imogen Denny then saw me through to the finishing line with patience and skill. My grateful thanks to all the team at Unbound – Georgia, Jimmy, Matthew and all – for their support with the early idea, the fund-raising, the book's website, uploading the interviews and

Acknowledgements

all the rest. Sincere appreciation to Lindeth Vasey for her excellent copyedit, and to Mark Bowsher for making the wonderful short film (set to Haydn's *Creation*) which we used to promote the book.

Finally, I would like to reiterate my heartfelt thanks to all those who supported the book and made it possible, including my patrons: Richard Davey (Dad), Hylton Murray-Philipson, Jessica and Adam Sweidan, Ben Goldsmith and Steve Wake. It meant so much to me to know that the project of the book, and the ideas behind it, were backed by the over 160 people who have so far contributed to the book's publication: family, dear personal friends, colleagues and people I do not know but who also got behind the project.

Thank you so much for supporting the book, and I do hope that the result is worthy of your support.

Unbound is the world's first crowdfunding publisher, established in 2011.

We believe that wonderful things can happen when you clear a path for people who share a passion. That's why we've built a platform that brings together readers and authors to crowdfund books they believe in – and give fresh ideas that don't fit the traditional mould the chance they deserve.

This book is in your hands because readers made it possible. Everyone who pledged their support is listed below. Join them by visiting unbound.com and supporting a book today.

Diana Angelo
Will Ashley-Cantello
James Atkins
Edwina Attlee
Sharon Bakar
Scarlett Benson
David Bent-Hazelwood
Anthony Blakeney

Becky Bolton
Mark Bowsher
Claire Bradbury
Fr Stephen Brown
Charles Budd
Wesley Burden
David Cadman
Ben Caldecott

Anna Chapman
Peter Cousins
Robert Cox
Tim Croker
Anna da Costa
Geoffrey Darnton
Benedict Davey
Cathy Davey
Nick Davey
Tom Davey
Alexandra de Laszlo
Davina de Laszlo
The de Laszlo Foundation
Amandine de Schaetzen
Caroline Donnelly
Stuart Elliott
Anna Emerson
Selin Erzin
Alex Evans
Rob Eynon
Seb Falk
Sonia Faruqi
Virginia Fassnidge
Bernadette Fischler
Katie Flanagan
Douglas Flynn
Alice Forrester
Rosa Fürstenberg
Leonardo Garrido
Lisa Genasci

Jack Gibbs
Global Counsel
Joanna Glynn
Benjamin Goldsmith
Chris Golightly
John Gollifer
Dominic Gould
Dermot Grenham
Martyn Griffiths
L Guidi
Jan Hall
Barrie Hargrove
Kimberly Harrington
Tom Harrison
Caitlin Harvey
Tom Hay
Alberto Heredia
Roger Hollies
Mary Horlock
Jeff Horne
Dawn Howley
Jeff Hutner
IDEAA Regeneration
   Systems
Minni Jain
Anni Kelsey
Ros Kennedy
Mark Keogan
Shamir Khona
Dan Kieran

Supporters

Ed King
Hugh Knowles
Roman Krznaric
Noelle Kumpel
Felipe Kuzuhara
Valerie Langfield
Alison Layland
Alice Lépissier
Leigh Linabury
Erynn Linabury and James
    Scott
Beatriz Luraschi
Stephanie Maier
Henry Mance
Jaime Martinez Bowness
John Peter Maughan
Vera Mazzara
Helen McCabe
Richard McColl
Rebecca McEnery
William McFarland
Josephine McGowan
Jenny McInnes
Paul McMahon
Paul Meins
David Mitchell
John Mitchinson
Katherine Montesinos
Zaki Moosa

Richard Morris
Cat Moyle
Justin Mundy
Carlo Navato
Katharine Norbury
David Nussbaum
Mayowa Ochere
Georgia Odd
Gopal D. Patel
Sergio Perez Leon
Adrian Phillips
Anthony Phillips
Barnaby Phillips
Oliver Phillips
Eduardo Plastino
Leo Pollak
Justin Pollard
Laura Pollard
Trishan Ponnamperuma
Jonathon Porritt
Toby Porter
Mirlo PositiveNature
Stuart Proffitt
Jimena Puyana
Jose R. Puyana
Kate Raworth
Magnolia Restrepo
Thinley Rhodes
Nick Robins

Caroline Robinson
Mark Russell (Avatar
  Wizard)
Adam Salt
Richard Scobey
Joana Setzer
Mike Shanahan
Gill Shepherd
Melanie Siggs
Luis Solorzano
Jeremy Spurway
St Edward's School Library
Jamie Stewart
Jung-ui Sul
David Taylor
Sheila Taylor
Kristian Teleki
Tom Thornley

Vivian, Thrive, Battersea
  Park
Paul Tompsett
Solitaire Townsend
Kate Upshon
Maurice van Beers
Elizabeth Venkatesham
Daniel Vennard
Nick Walpole
Loosi Williamson
Florence Wolfe
Michael Wolosin
Patrick Wylie
Mohammed Zafir
Benjamin Zilker
Sarah, Barry
  & Gabby Zins